Sustainable Aviation

Series Editors

T. Hikmet Karakoc ⓘ, Faculty of Aeronautics and Astronautics, Eskisehir Technical University, Eskisehir, Turkey; Information Technology Research and Application Center, Istanbul Commerce University, Istanbul, Turkey

C Ozgur Colpan ⓘ, Department of Mechanical Engineering, Dokuz Eylül University, Buca, Izmir, Türkiye

Alper Dalkiran ⓘ, School of Aviation, Süleyman Demirel University, Isparta, Türkiye

The Sustainable Aviation book series focuses on sustainability in aviation, considering all aspects of the field. The books are developed in partnership with the International Sustainable Aviation Research Society (SARES). They include contributed volumes comprising select contributions to international symposiums and conferences, monographs, and professional books focused on all aspects of sustainable aviation. The series aims at publishing state-of-the-art research and development in areas including, but not limited to:

- Green and renewable energy resources and aviation technologies
- Aircraft engine, control systems, production, storage, efficiency, and planning
- Exploring the potential of integrating renewables within airports
- Sustainable infrastructure development under a changing climate
- Training and awareness facilities with aviation sector and social levels
- Teaching and professional development in renewable energy technologies and sustainability

* * *

Sergii Boichenko • Anna Yakovlieva
Oleksandr Zaporozhets • T. Hikmet Karakoc
Iryna Shkilniuk • Alper Dalkiran
Editors

Sustainable Transport and Environmental Safety in Aviation

Editors
Sergii Boichenko
National Technical University of Ukraine
"Igor Sikorsky Kyiv Polytechnic Institute"
Kyiv, Ukraine

Anna Yakovlieva
Technical University of Kosice
Kosice, Slovakia

Oleksandr Zaporozhets
Łukasiewicz Research Network-Institute
of Aviation
Warsaw, Poland

T. Hikmet Karakoc (iD)
Faculty of Aeronautics and Astronautics
Eskisehir Technical University
Eskisehir, Turkey

Iryna Shkilniuk
National Technical University of Ukraine
"Igor Sikorsky Kyiv Polytechnic Institute"
Kyiv, Ukraine

Information Technology Research
and Application Center
Istanbul Commerce University
Istanbul, Turkey

Alper Dalkiran (iD)
School of Aviation
Süleyman Demirel University
Keciborlu, Isparta, Türkiye

ISSN 2730-7778 ISSN 2730-7786 (electronic)
Sustainable Aviation
ISBN 978-3-031-34349-0 ISBN 978-3-031-34350-6 (eBook)
https://doi.org/10.1007/978-3-031-34350-6

© The Editor(s) (if applicable) and The Author(s), under exclusive license to Springer Nature Switzerland
AG 2023
This work is subject to copyright. All rights are solely and exclusively licensed by the Publisher, whether
the whole or part of the material is concerned, specifically the rights of translation, reprinting, reuse of
illustrations, recitation, broadcasting, reproduction on microfilms or in any other physical way, and
transmission or information storage and retrieval, electronic adaptation, computer software, or by
similar or dissimilar methodology now known or hereafter developed.
The use of general descriptive names, registered names, trademarks, service marks, etc. in this publication
does not imply, even in the absence of a specific statement, that such names are exempt from the relevant
protective laws and regulations and therefore free for general use.
The publisher, the authors, and the editors are safe to assume that the advice and information in this
book are believed to be true and accurate at the date of publication. Neither the publisher nor the authors or
the editors give a warranty, expressed or implied, with respect to the material contained herein or for any
errors or omissions that may have been made. The publisher remains neutral with regard to jurisdictional
claims in published maps and institutional affiliations.

This Springer imprint is published by the registered company Springer Nature Switzerland AG
The registered company address is: Gewerbestrasse 11, 6330 Cham, Switzerland

Preface

The book selectively represents the chemmotological aspects of sustainable transport development and providing environmental safety of modern aviation. It is devoted to the relevant problems of rational use of conventional and alternative fuels, lubricants, and other operation materials, as well as environmental and economic aspects of road and air transport exploitation. This book, *Sustainable Transport and Environmental Safety in Aviation*, is a second volume that comprises selected outstanding papers presented at the VIII International Scientific-Technical Conference, "Problems of Chemmotology: Theory and Practice of Rational Use of Traditional and Alternative Fuels and Lubricants." This event was held in Kamianets-Podilskyi, Ukraine, from June 21 to June 25, 2021. The conference is a traditional event, which gathers researchers, scientists, practicians, and academics in the field of chemmotology, aviation, transport engineering, environmental safety, recycling and utilization, and sustainable development from all over the world. Traditionally, the main aim of the conference is meeting of qualified specialists capable of solving tasks of any level of complexity in the field of chemmotology, exchange of experience, information, discussion of issues and options for their solution, development of intellectual potential for knowledge management with the aim of turning it into intellectual capital, ensuring the most effective results through a holistic and adequate approach to solving modern problems in the key conference's fields, integration of the experience of the older generation and young scientists, as well as the support of qualitative research and development of new technologies aimed at increasing the rational use of fuels and lubricants, technical liquids, and additives.

The book is composed of nine chapters in total. All chapters presented by the authors (co-authors) are published in the author's edition and aim to present an issue on how to achieve more sustainable and more environmentally safe development of modern transport. The first chapter is devoted to the implementation of information management for air safety provision. The second chapter includes recent studies in the fields of information analysis and control of its reliability. The third chapter of the book includes the results of the studies on the environmental impact assessment of air transport activity. The fourth chapter presents issues related to sustainable recycling and the utilization of waste batteries. The fifth chapter analyzes aspects of information protection in electric transportation systems. The sixth chapter covers aspects of environmentally friendly applications and the promotion of novel chemical technologies. The seventh chapter presents the environmental and marketing aspects of the vortex granulators' application in chemical engineering. The eighth chapter is devoted to the environmental and economic assessment of automobile road construction. And, the ninth chapter covers the issues of reducing negative environmental impact of transport engines by implementing new hydrogen technologies.

This book will be interesting and useful for the professional career of operators of air transport, jet fuels suppliers, professionals in the sphere of transport environmental safety, alternative fuel manufacturers at oil processing plants, fuels, lubricants and fuel additives producers, aviation fuel handling companies, etc. The book will be also interesting for researchers at all stages of careers.

The contributions of the authors and reviewers and the assistance of conference organizers in the preparation of this book are sincerely appreciated.

Kyiv, Ukraine	Sergii Boichenko
Kosice, Slovakia	Anna Yakovlieva
Warsaw, Poland	Oleksandr Zaporozhets
Eskisehir, Turkiye	T. Hikmet Karakoc
Kyiv, Ukraine	Iryna Shkilniuk
Keciborlu, Isparta, Turkiye	Alper Dalkiran

Contents

1 Air Safety Information Management 1
Pavol Kurdel, Marek Češkovič, Alena Novák Sedláčková,
and Jaroslav Zaremba

**2 Increasing the Reliability of Diagnosis and Control
in the Uncertainty of Primary Information** 13
Pavlo Schapov, Olga Ivanets, Pavlo Kulakov, and Larysa Kosheva

**3 Environmental Impact Assessment of the Planned
Activity of Aviation Transport** 37
Viktoriia Khrutba, Tetiana Morozova, Anna Kharchenko,
Inesa Rutkovska, and Alla Herasymenko

**4 Key Aspects of Sustainable Development Toward Spent
Lithium-Ion Battery Recycling** 59
Lina Kieush, Andrii Koveria, Andrii Hrubiak, and Serhii Fedorov

**5 Green Technologies of Information Protection in Computer
Networks of Electric Transport System** 75
Halyna Holub, Ivan Kulbovskyi, Vitalii Kharuta, Olga Zaiats,
Mykola Tkachuk, and Valentyna Kharuta

**6 Multistage Drying in Fluidized Bed: Ways of Eco-friendly
Application and Marketing Tools for Promotion** 91
Nadiia Artyukhova, Tetiana Vasylieva, Serhiy Lyeonov,
Jan Krmela, Oleksandr Shandyba, and Olena Melnyk

**7 Vortex Granulators in Chemical Engineering: Environmental
Aspects and Marketing Strategy of Implementation** 107
Artem Artyukhov, Nadiia Artyukhova, Jan Krmela,
Tetiana Vasylieva, Serhiy Lyeonov, and Olena Melnyk

viii

8 Evaluation of Automobile Road Construction Environmental and Economic Efficiency Based on Public-Private Partnership 123
Yevheniia Tsiuman, Mykola Tsiuman, and Anatolii Morozov

9 Improving the Energy Efficiency and Environmental Performance of Vehicular Engine Equipped Within the On-Board Hydrogen Production System 143
Mykola Tsiuman, Vasyl Mateichyk, Miroslaw Smieszek, Ivan Sadovnyk, Roman Artemenko, and Yevheniia Tsiuman

Index 167

Chapter 1
Air Safety Information Management

Pavol Kurdel, Marek Češkovič, Alena Novák Sedláčková, and Jaroslav Zaremba

Nomenclature

FO	Flying objects
NEC	Normal exploitation conditions
DFC	Difficult flight conditions
DFS	Difficult flight situation
ES	Emergency situation
CS	Catastrophic situation

1.1 Introduction

The functionality of the aviation ergatic complex is determined by the quality of the defined relationship between the two subjects – *aircraft* and *human*. The safe control of flying apparatus (aircraft, autonomous aircraft, etc.) is a guideline for the action of the independent parameters of their onboard systems, which are reflected in the perceived sense of security and safety of the pilot or operator of the autonomous aircraft. Fulfilling the safety manifestations of flying devices requires the implementation of aircraft control in way that the flight will take place in a credible flight

---

P. Kurdel (✉) · M. Češkovič
Technical University of Košice, Košice, Slovakia
e-mail: Pavol.kurdel@tuke.sk; marek.ceskovic@tuke.sk

A. N. Sedláčková
University of Žilina, Žilina, Slovakia
e-mail: alena.sedlackova@fpedas.uniza.sk

J. Zaremba
The University of Security Management in Košice, Košice, Slovakia

© The Author(s), under exclusive license to Springer Nature Switzerland AG 2023
S. Boichenko et al. (eds.), *Sustainable Transport and Environmental Safety in Aviation*, Sustainable Aviation, https://doi.org/10.1007/978-3-031-34350-6_1

environment. The implemented safe flight can be achieved if the chosen methods and procedures of control of the flying apparatus will reflect real processes in the monitored safe flight environment (Lazar et al. 2011). The peculiarity of flight safety monitoring is a quantitative estimation in individual flight modes with the simultaneous action of unforeseen harmful factors, including the psychophysiological capabilities of the crew, which is also the subject of research of numerous scientific teams. Quantitative estimation of the level of safety of flying objects (FO) in individual periods of operation is performed according to statistical and probabilistic criteria and indicators (Žihla et al. 1988). The statistical criteria for quantitative estimation are used to calculate aviation safety for the entire history of the existence of a particular type of aircraft or its operating period. General statistical indicators of the safety of flying objects are as follows:

- $n_{aap}(T)$ – number of air accident predictions of flying object (FO) in a certain time period T
- n_{na} – number of accident
- n_{nad} – number of air disasters
- n_d –number of deaths in an air accident in the observed period m(T)
- a_e – aviation event.

The number of unforeseen aviation events is also an absolute indicator of safety (FO) a_e caused j-th factor and the number of unforeseen events a_e on i-th of flight period. Statistical indicators are important for evaluating safety (FO) and its trends. Their shortcoming is that they do not reflect the current level of aviation safety or the amount of effort made by air carriers. Proportional indicators seem to be more suitable: the cost of the period of flight time in air transport, the volume of work performed (L) per flight, the number of transported passengers, cargo, etc. (Taran 1976). The position of the universality of statistical indicators in the flight period is characterized by relative safety parameters (FO). The increase in aviation events depends on human and technical factors. Their manifestation is more frequent during long flights (L), or from the count of technical failures that arose on the ground during the technical life of flight objects (aircraft, UAV etc.). It presents the formula:

$$kf_j = nf_1/Lf \tag{1.1}$$

where nf_1 – number of damage to aircraft (afc) on the ground,

f_1- index; flight event, type of aircraft (gradually).

When dealing with flight incidents, the valid regulations of the International Civil Aviation Organization (ICAO) must be followed. This organization uses the following relative statistics in its safety sections:

1 Air Safety Information Management

$$kl = \frac{nk}{L10^{-8}}; kT = \frac{nk}{T_1 10^{-5}}; kN = \frac{nk}{N10^{-5}}; kl_1 = \frac{n}{A_{pax}10^{-6}}; kl_2 = \frac{m}{A_n}$$
$$- k10^{-8} \tag{1.2}$$

where

k_L, k_T, k_N – number of air disasters per 100 mil. Kilometers, for one hundred thousand raids per hour, respectively one hundred thousand flights (gradually),

L, T_1, N – flight distance in kilometers, an flight time in hours and number of flights in the analyzed period,

kl_1, kl_2 – the number of dead passengers with impact per one million passengers carried and per 100 million passenger-kilometers in the analyzed period.

1.2 Formalization and Methods for Solving the Quantitative Evaluation of the Safety of the Ergatic Complex

Due to the effects of random flight influences and unwanted factors that interact both positively and negatively, each flight is a result of coincidence (Kozaruk). The mathematical method used to model random events in safety analysis (FO) accepts probabilistic criteria. For acceptable safety (FO), the following designation P (probability of safe flight) will be used; for a dangerous flight, the probability is $Q = 1 - P$.

The determining factor of the safety criterion is the number of identical (where safety conditions are the same) flights performed (Schimidt 1996). The required criterion is also the frequency and independence of events. In this case, the criteria may be subject to the Poisson distribution of probability. When during n – flights (FO) have happened n_h – hazards and assumptions of flight accidents:

$$Qn_{Lh}^N = \left[\frac{(NQ)^{nLh}}{nLh!}\right]e^{-NQ} \tag{1.3}$$

The probability of safely completing all of n – flights will be determined by the following formula:

$$PN = Q_0; PN = e^{-NQ} = e^{-(N(1-P))}; \tag{1.4}$$

The probability Q_n^{Lh} represents a complex function depending on the properties of the aeronautical ergatic complex, specifically in its content such as the following:

X, which represents the vector of parameters of the movement of the flying object
W, which presents the properties of external influences in the flight environment
Z, description of the properties of the ergatic aviation transport complex
R, all possible external influences of unwanted parameters that will cause nonstandard operating conditions of the air transport system (Schimidt 1996)

The characteristics of probable safety upon acceptance (1.3) and (1.4) and the following factors will take on a mathematical formula:

$$P_{(t)} = 1 - q_t = F\,[X_t, W_t, Z_t, R_t] \tag{1.5}$$

Equation 1.5 represents a syllabus of research and problem-solving in flight safety theory.

1.3 Methods for Solving Safety Problems According to the Theory of Probability

The current results of the research are mainly focused on selected problems of flight safety with an orientation on failures of aircraft and errors of aviation personnel. Equation 1.5, as a formal statement, highlights and determines the conditions for the implementation of the selected part of security (Jadlovská and Jadlovská 2013). This is placed in the quality and quantity of security assurance of the functions of the system elements that make up the content of information security (5). Complex aeronautical systems are subject to controlled safety effectiveness criteria in various forms. The general criterion is normalized to formulas and vectors during implementation:

$$J_i = J_i/a_{im}; J_i = (J_i - a_{im})/(a_{im} - a_i); \tag{1.6}$$

where

a_{im}, a_i is selected maximum and minimum standard models of possible (gradually) criterion values J_i. The information function is a combination of normalized expressions (1.6) into the following expression:

$$\Phi\,(\alpha_1, J_1 \ldots \ldots \alpha_{im}, J_m); \tag{1.7}$$

where we require the sum of the coefficient of severity to be $\alpha_i = 1$.

By the standards and rules for transport aircrafts, the probabilities of special situations were determined H_i.

1.4 Construction of the Informative Function of Dangerous Faults on Aviation Objects

The use of information control systems for the safety of the operation of airborne ergatic systems of an aircraft object requires the acceptance of the outputs of control diagnostic devices and devices capable of identifying faults. The flight crew needs solutions if they cannot identify failures in a non-standard flight situation. In such situations, currently, on-board safety systems can help the crew with automatic activation of the safety modes of the aircraft systems (Schimidt 1996). In accordance with the type of the aircraft and its shape (aerodynamic), the critical parameters are defined by standards that are quantified for each flight mode (FO) (Kurdel 2016). The identification of the condition is then possible through the integral informative functions of the hazard and the characteristic data of the special situations, $- x_{Hi}$; characteristic situations at the flight period, x_{ni}; permissible flight situation, x_{per}; and critical situation, x_{crit}. Standards such as the appropriateness of air transport aircraft are assigned to these characteristics, with the value of the special situation's method as per Fig. 1.1. This means that each special flight situation Zi belongs to the critical parameter x_i according to which it can be identified by inequality:

According to Table 1.1, it is possible to illustrate special flight situations with models (1.3) and (1.4), which in natural logarithmic form in the Matlab model environment are as follows:

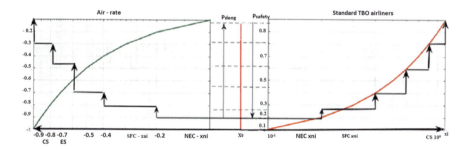

Fig. 1.1 Graphical interpretation of standardized parameters in special flight situations

Table 1.1 Models of special flight situations

(a)	(b)	(c)
x_0	Normal exploitation conditions "NEC"	$10^{-9} < P_{NEC} < 10^{-6}$,
xu_C^{DFCi}	Difficult flight conditions "DFC"	$10^{-6} < P_S^{FC} < 10^{-4}$,
xs_S^{Sni}	Difficult flight situation "DFS"	$10^{-4} < P_S^{FS} < 10^{-2}$,
x_E^{Sni}	Emergency situation "HS"	$10^{-2} < P_E^S < 1$,
x_C^{Sni}	Catastrophic situation "CS"	CS $=1$;

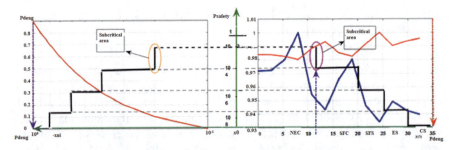

Fig. 1.2 Standard probabilistic parameters of airworthiness safety of transport aircraft

> xiHi=-0.1:-.1:-1;,
>
> subplot(2,2,1),semilogx(-xiHi,-xiHi),grid on,
>
> subplot(2,2,2),semilogx(xiHi,-xiHi),grid on,
>
> xiPsafety=0.1:0.1:1;
>
> subplot(2,2,3),semilogx(-xiPsafety,-xiPsafety),grid on,
>
> subplot(2,2,4),semilogx(xiPsafety,-xiPsafety),grid on,

According to characteristic values in Table 1.1b, it can be generalized that values x_{ni} transform into operating parameters with immediate *special*, *permitted*, and *critical* values:

$$x_{Hi}, \; x_{ni}, x_{po}^{ni} \; a \; x_{crit}^{ni}.$$

The airworthiness standards of transport aircraft are set to appropriate values x_i affiliate according to the inequalities listed in Table 1.1b. At the limits of each normalized value, the limit critical lines correspond in value to the special situations according to Eq. 1.8:

$$x_i = Q_n^{fe}, N \tag{1.8}$$

The display of values is on Fig. 1.2.

Similar to the borders of special situations Z_i, values of critical parameters for the probability of flight mode hazard indicators P_{xi}^{le} are unambiguous. According to the limit x_i^{zi} points, it is possible to construct the continuous information function determined along the following:

$$\Phi = f(x_i) \sim P_{xi}^{le}, = f(x_i) \sim P_{xi} \, xi^{le} \tag{1.9}$$

1 Air Safety Information Management

Equation 1.9 characterizes the level of flight safety (danger) by change of critical parameter x_i (see Fig. 1.2).

An example of the implementation of the described methodology can be a model of special solutions for sorting the aircraft descent according to the known parameters of the trajectory for final approach and landing (Džunda et al. 2018; Džunda and Kotianová 2016).

The peculiarity of the regime is, for example, longitudinal gusts of wind. Integral models are Eqs. 1.3, 1.4, and 1.5. In their implementation, we will use models of informative functions (Kozaruk and Rebo 1986):

$$F_{\text{deng}}^{\alpha} = \left\{ (\alpha - \alpha_{\text{per}}) / (\alpha_{\text{crit}} - \alpha_{\text{per}} \right\}^{1} / n_1 \qquad (1.10)$$

$$F_{\text{deng}}^{as} = \left\{ (V_{\text{per}} - V) / (V_{\text{per}} - V_{\text{crit}} \right\}^{n_2} \qquad (1.11)$$

where

F_{haz}^{α}– *informative function of danger according to the alpha (angle of attack)*
F_{haz}^{As}– *informative function of danger according to airspeed (as): crit, critical values per – permitted values. n_1, n_2– reciprocity of the angle of attack and airspeed*

The links are $n_1 = 2$. Link F_{deng}^{as} is defined by the following relation:

$$n_2 = \left\{ (\alpha_{\text{crit}} - \alpha_{\text{per}}) / (\alpha_{\text{crit}} - \alpha) \right\} \qquad (1.12)$$

The informative safety function is defined by the Weibull density of the distribution (5):

In accordance with (1.10), (1.11), and (1.12), the informative function of danger of the aircraft positioned according to the descent plane when landing will be determined by the following expression:

$$F_{\text{deng}} = 1 - \left\{ 1 - \left(F_{\text{deng}}^{\alpha} \right)^{1/2} \right\} \cdot \left\{ 1 - \left(F_{\text{deng}}^{as} \right)^{((\alpha_{\text{crit}} - \alpha_{\text{crit}}) / (\alpha_{\text{crit}} - \alpha))} \right\} \qquad (1.13)$$

The informative hazard function (1.13) changes when α and *as* change in the range from 0 to 1. This means that in the area of changes α, Airsp, it is possible to record several points with an unchanged value F_{deng}, line of safety. The shape of the lines of the same safety depends on the degree of the approach of the controlled critical parameters to the permitted limits.

1.5 Data Information Functions of Safety When the Aircraft Descends on Glide Path

In the given implementation, we are simulating the danger function of the flight of the aircraft on the descent – glide path in the Matlab environment:

syms As alfacrit alfaper alfa Asper Ascrit ni,

alfa=alfacrit; Then:

Fhaz=1-(1-((alfa-alfaper)./(alfacrit-alfaper)).^0.5).*(1-((Ascrit-As)./(Asper-Askcrit)).^(alfacrit-alfa)./(alfacrit-alfaper));

Then:

Fhaz1 =1;for all AS. Where:

Ascrit=As; will be:

Fhaz2=1-(1-((alfa-alfaper)./(alfacrit-alfaper)).^0.5).*(1-((Ascrit-As)./(Asper-Ascrit)).^(alfacrit-alfa)./(alfacrit-alfaper));

Fhaz2 =1;for all alfa.

Let the safety automata react to the unusual position of the aircraft's control systems, which caused a change in the alpha angle of attack and the flight speed (Kurdel 2016):

alfa=0:.1:30;As=0:.3:90; Critical and allowable values are:

alfacrit=40;degree,

alfaper=30;degree,

Ascrit=65;m/s;

Asper=85;m/s;

Then (12) je:

Fhaz=1-[1-((alfa-30)./10).^0.5].*[1-((85-As)./20).^(10./(40-alfa))];

X=real(Fhaz);

Y=imag(Fhaz);

absFbhaz=(X.^2+Y.^2).^0.5;

Standardization:

absFbhazN=absFbhaz./1.6221;

Fhaz=1-absFbhazN;

Safety:

Fhaz=1-Fhaz;

tlg1=0:.1:30;tlg2=0:.3:90;

figure(1);plot(tlg1,Fsaf,'b','LineWidth',3),grid on,hold on,

figure(1);plot(tlg1,Fdeng,'r','LineWidth',3),grid on,

title('Glisade flight information functions','FontSize',14),

ylabel('Informative function values','FontSize',14),

xlabel ('Time period stick control positioning and speed control','FontSize',14),

1 Air Safety Information Management

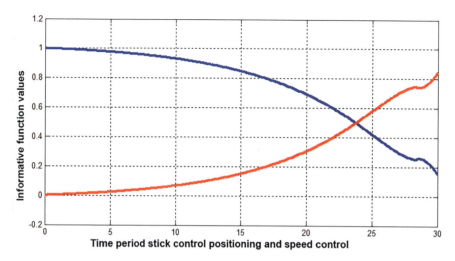

Fig. 1.3 Informative functions of flight safety during glide path descent

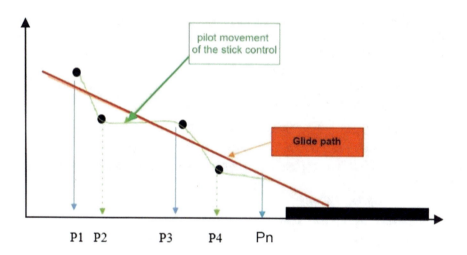

Fig. 1.4 Tracked waypoints during aircraft descent and their derivatives

In real practice, the gradient of the informative danger function is the result of finding and determining the rate of change of functional values (angle of attack alpha and airspeed) in the vicinity of the descent plane, in the direction of the given descent vector (Novak et al. 2019). As can be seen, this rate of change is exactly the value of the derivatives in the direction of a given descent vector (Fig. 1.3). In this context, the gradient itself has a geometric and physical meaning for the interpretation by the derivatives at these points ($P1 \div Pn$) in Fig. 1.4.

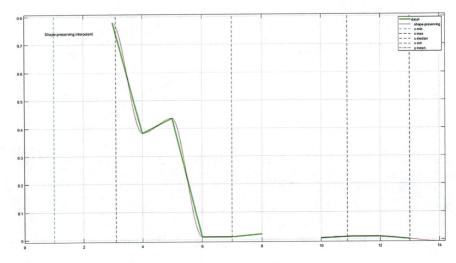

Fig. 1.5 The gradient-informative hazard

We can find the maximum and minimum values during the descent when piloting the aircraft in semiautomatic descent mode (without autopilot) according to the navigation points (Fig. 1.5).

syms alfa As Fdeng

alfa=[3.8 4 3.8 3.9 4.15 4.14 4.16 4.18 4.19 4.18 4.17 4.16 4.18];

As=[20.7 20.69 20.55 20.27 21.11 21.5 20.83 20.55 21.38 21.94 21.27 21.52 21.66];

Fdeng=[1-(1-((alfa-3.5)./(4.19-3.5)).^0.5).(1-((20-Vv)./(19-21)).^(4.19-alfa))];*

Fdeng =[0.8856   0.9728 0.8653 0.8949 0.9993 0.9995 0.9994 0.9999 1.0000 1.0000 0.9999 0.9998 1.0000];

gradient(Fhaz, [alfa, As]);

plot(ans,'g','LineWidth',3),grid on,

title('Gradient of the informative function of danger','m','FontSize',14),

1.6 Conclusion

The performed analysis is in accordance with the actually valid rules and the directions for air traffic. The resulting effect is in the methodology of the design of pedagogical research and its direction to air transport ergatic systems. The scientific and technical character of the subject of research is concretized, graphically based and implemented so that it can be used both in pedagogical and aviation practice.

The used methodology of analysis and synthesis (Figs. 1.1, 1.2 and 1.3) shows the principle of creating information functions and its shift through precritical areas to ongoing flight situations. The information for the procedure is performed by intelligent sensor systems. The used mathematical models of flight situations in the form of the presented equations accepted the basic design and aerodynamic properties of transport aircraft.

The special contribution of the authors is as follows:

- The analyzed theoretical methods ensure the possibility of evaluating the current flight safety problems.
- Methodology of shaping informative functions in semilogarithmic coordinates.
- Methodology of selection and placement of x_{ni} sensors in aircraft situation state.

References

Džunda M, Kotianová N (2016) The accuracy of relative navigation system production management and engineering sciences – scientific publication of the international conference on engineering science and production management. ESPM 2015:369–376

Džunda M, Dzurovcin P, Cekanova D (2018) The model of flying objects in the relative navigation system. In: Transport means – proceedings of the international conference, pp 1050–1055

Jadlovská A, Jadlovská S (2013) Moderné metódy modelovania a riadenia nelineárnych systémov (Modern methods of modeling and control of nonlinear systems). Elfa s.r.o., Košice

Kozaruk VV, Rebo JJ (1986) Navigacionnye ergatičeskie komplexy samoletov. Moskva, Mašinostrojenie, 288 stran

Kurdel P (2016) Vedecko-pedagogické aspekty experiment a modelovania so zložitými leteckými systémami (Scientific and pedagogic aspect of experiment and modelling with complex aviation systems). TUKE, Košice, Habilitačná práca

Lazar T, Bréda R, Kurdel P (2011) Inštrumenty istenia letovej bespečnosti. Košice. 232 stran, Vysokoškolská učebnica

Novak A, Skultety F, Bugaj M et al (2019) Safety studies on gnss instrument approach at caronilina airport. In: MOSATT 2019 – modern safety technologies in transportation international scientific conference, proceedings, pp 122–125

Schimidt DT (1996) Cooperative synthesis of control and display augmentation, Wiliamsburg, Colect. Technical paper. New York, pp 732–742

Taran VA (1976) Ergatičeskie sistemy upravlenija. Ocenky kačestva ergatičeskich procesov, Moskva: Mašinostrojenie, 189 stran

Žihla Z, Jalovecký J, Pařízek R (1988) Automatické řízení letadel. Časť 2. Brno, 263 stran

Chapter 2
Increasing the Reliability of Diagnosis and Control in the Uncertainty of Primary Information

Pavlo Schapov, Olga Ivanets, Pavlo Kulakov, and Larysa Kosheva

Nomenclature

CO	Carbon monoxide
CO_2	Carbon dioxide
CH_4	Methane
C_2H_2	Acetylene
C_2H_4	Ethylene
C_2H_6	Ethane

2.1 Problem Resolution

Credibility is a basic measure of the quality of nondestructive testing. Increasing reliability is the task of reducing uncertainty in obtaining primary information about changes in the properties of an object of control or diagnostics.

The main sources of uncertainty of the primary information about the control parameter contained in the measurement results of the controlled quantities are the following factors:

P. Schapov
National Technical University "Kharkiv Polytechnic Institute", Kharkiv, Ukraine

O. Ivanets (✉) · L. Kosheva
National Aviation University, Kyiv, Ukraine
e-mail: kosheva@npp.nau.edu.ua

P. Kulakov
Uman National University of Horticulture, Uman, Ukraine

© The Author(s), under exclusive license to Springer Nature Switzerland AG 2023
S. Boichenko et al. (eds.), *Sustainable Transport and Environmental Safety in Aviation*, Sustainable Aviation, https://doi.org/10.1007/978-3-031-34350-6_2

- Errors of technical measurements carried out during control, on the basis of non-standardized measuring instruments, including noncertified nondestructive testing means (Shchapov and Avrunin 2011; Bondarenko et al. 2007)
- The complexity of the formalized description of the control object, especially taking into account the random effects of the external environment
- Restrictions on the collection of primary measurement information at the stage of training the monitoring and diagnostic system
- Nonstationarity of controlled values in time, especially for dynamic objects of control and diagnostics (Shchapov et al. 2015)

Despite the ambiguity of the listed factors, they are united by the fact that they exist as a whole within the framework of an unknown model of the stochastic influence of quantitative changes in control parameters on the probabilistic properties of the vector of controlled information signals. The theoretical substantiation of the choice of such a model, when creating control systems for stochastic parameters, should be carried out at the training stage, ensuring the minimum uncertainty of the results of multidimensional transformations at the stages of control or diagnostics, regardless of the type of influencing factors. It should be borne in mind that the choice of the dimension of the vector of monitored signals and the number of levels of the monitored parameter determine the reliability of decision-making and depend on the incompleteness of the primary information about the properties of the object of control and diagnostics obtained at the training stage. And although modern information technology is the largest branch of technical cybernetics, the methods and theoretical developments of the latter are not used effectively enough to improve information and measurement technologies for multi-alternative control and functional diagnostics of objects with random information signals.

Basically, these methods are aimed at solving particular problems of reducing the effect of factor influence on the completeness of measurement information.

2.2 Analysis of the Models of Controlled Quantities in Conditions of Statistical Inhomogeneities of the Measurement Experiment

2.2.1 General Model of Parameter Control

The implementation of the requirements for ensuring a given completeness of information in control systems encounters difficulties of a fundamental nature:

- Controlled physical characteristics (control parameters) can be so complex that even standard samples are not available for them.
- Levels of directly controlled physical quantities measured are tied to a priori uncertain states of a physical object of measurement control and cannot be reproduced unambiguously for any of the fixed states.

The physical complexity of the control parameters may be due to the requirements of their maximum information content in relation to qualitative changes in the control object, and such parameters, most often, are complex physical quantities that are inaccessible to direct measurements. As for the controlled quantities (or information parameters), for industrial control objects (technological lines, industrial equipment, mechanical systems, etc.), these indicators are random processes with different types of nonstationarity.

In any case, we are talking about control objects, for which the number of information parameters theoretically tends to infinity, and the levels of a finite number of control parameters cannot be reproduced with metrological accuracy for any of the normative states of the control object.

Reducing the number of control parameters due to the impossibility of ensuring the accuracy of measurements of their levels leads to the loss of a significant part of the information. This is expressed in the fact that the parameters, which are not informative in one, for example, the normal state of the control object, acquire special significance in other states. It became necessary to plan the stages of collection and processing of primary information with complex relationships reflecting the structure of the controlled object at the level of sophisticated mathematical models of information transformation.

Such models can serve as the basis for a normative approach in assessing the qualitative state of a controlled object as a multifactorial object, for which any measurement experiment should include mathematical planning methods under conditions of uncontrolled factor influences (or inhomogeneities) that distort the results of such an experiment.

During technological measuring control, for example, the sources of discrete type heterogeneities can be differences in the composition of raw materials, elements of technological equipment, methods of processing primary information, etc. The nature of discrete sources of inhomogeneities is qualitative, leading to random errors in the measurement experiment.

Sources of continuous-type inhomogeneities are gradual parametric changes in the properties of the controlled object, leading to distortion of the measurement information signals in the form of drift or nonstationarity of the mean value, dispersion, and spectral components. Most often this is caused by the aging of elements of technological equipment or elements of measuring instruments. The nature of sources of continuous inhomogeneities is quantitative, leading to nonrandom systematic errors in the measurement experiment.

Regardless of the type of inhomogeneities, we can talk about unavoidable influencing factors, which, however, can be taken into account when planning measuring experiments, both at the stage of training the measuring control and diagnostics system and at the stages of practical use of this system as part of quality assurance management during industrial production of products and operation of complex technical systems.

All information available to the researcher about changes in the state of the controlled object is contained in the vector of controlled values $\overline{X}$, the components of which are available for direct measurements (single or multiple).

Let x_{li} be the result of the i-th measurement of the X_l component $l = \overline{1, N_l}$ at time t_i (N_l is the number of measurements for the l-th component). The general model of the stochastic influence of the control parameter Y and the set $\{X_v\}^{n-1}$, $v \neq l$, of physical indicators on the input value X_1 has the form

$$X_l(t/Y) = F_l[Y, t] + \varepsilon_l\left[Y, \{X_v\}^{n-1}, t\right] + \varepsilon_l(t) \tag{2.1}$$

where:

$F_l[\cdot]$ is the main informational component of model (2.1)
$\varepsilon_l[\cdot]$ is additional informational component
$\varepsilon_1(t)$ is residual random component

The second component $\varepsilon_l[\cdot]$ can be used to simulate discrete inhomogeneities that transform this component into a discrete random variable functionally dependent on the observation time t.

The main information component $F_l[\cdot]$ can simulate continuous irregularities in the form of an uncontrolled drift of the vector Y in time t.

In the future, we will consider three models of the uncertainty of the measurement results of controlled quantities, determined by the type of inhomogeneity of the measurement experiment:

1. Static quantity with discrete-type inhomogeneities

$$X_l(Y) = F_l[Y] + \varepsilon_l\left[\{X_v\}^{n-1}\right] + \varepsilon_l \tag{2.2}$$

2. Dynamic quantity with discrete-type inhomogeneities

$$X_l(t/Y) = F_l[Y] + \varepsilon_l\left[\{X_v\}^{n-1}, t\right] + \varepsilon_l \tag{2.3}$$

3. Dynamic quantity with inhomogeneities of continuous and discrete types

$$X_l(t/Y) = F_l[Y, t] + \varepsilon_l\left[\{X_v\}^{n-1}, t\right] + \varepsilon_l \tag{2.4}$$

In the general model (2.1), the third component $\varepsilon_l(t)$ reflects the instrumental error of technical means of control. The second component reflects the uncertainty due to the factor influence of Y and $\{X_v\}^{n-1}$ on the information parameters X_l. The first component models the influence of the transformation function, the type and parameters of which are also uncertain, but can be replaced by a functional model, the degree of inadequacy of which also generates uncertainty within the systematic error (Shchapov et al. 2015).

2.2.2 Mathematical Model of a Static Controlled Quantity with Discrete-Type Inhomogeneities

By the object of control, we mean a material (physical) object that has numerous features (physical properties) that are in multifaceted and complex interrelationships, and one or more of these properties can be measured.

By the model of an object of control, we mean a mathematical description that reflects both the characteristics of the object and the uncertainty of the assessment of its qualitative state based on the results of measuring the controlled quantities that characterize individual properties of the object.

Such a model should provide the maximum possible amount of primary information in relation to the two tasks solved during the creation of the control system:

1. Analysis of the probabilistic structure of the control object (training stage)
2. Synthesis of the optimal multi-parameter control system (control and diagnosis stage), where the optimality criterion is the maximum reliability of decision-making regarding the state of the control object.

Each of the tasks can be implemented within its own version of model (2.2) of controlled quantities.

2.2.3 Controlled Variable Model at the Training Stage

The physical object of control is represented by a set $\pi = \{\pi_1, \ldots \pi_k\}$ of its states (levels of parameter Y); for each of which there are n_j, $j = \overline{1,k}$, samples for the vector $\overline{X}$. The total number of vector values presented for analysis $\overline{X}$ is $N = \sum_{j=1}^{k} n_j$.

The technical means of control carry out direct measurements of the quantities $X_1, \ldots X_n$, presenting the values of these quantities in the form of point $\overline{X}^*$ vector estimates $\overline{X}$. According to the results of multiple N measurements, a multivariate statistical analysis of sample values of the vector $\overline{X}$ is carried out, which allows:

(a) To choose a model of the conditional n-dimensional distribution density $f\left(\overline{X}/Y\right)$ of the vector $\overline{X}$
(b) To rank the components of the vector of controlled quantities $\overline{X}$ in terms of information content with respect to changes in the levels $y_1, \ldots .y_k$ of the parameter Y
(c) To minimize the number of vector components $\overline{X}$ while maintaining the maximum value of the expected information about the values of the levels $y_1, \ldots .y_k$ of the control parameter Y, with a minimum of measurement uncertainty

The listed tasks can be solved within the framework of multivariate analysis of variance (Ivanets and Morozova 2021) of the results of multiple measurements of the

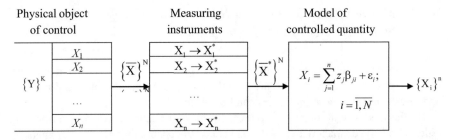

Fig. 2.1 Block diagram of the process of processing the results of multiple measurements at the stage of analysis of training samples. (Shchapov et al. 2015)

components $X_1,\ldots X_n$ under the conditions $Y \in y_j$, $j = \overline{1,k}$ and the a priori assumption of a nondegenerate normal distribution of the vector of control indicators $\overline{X}$ for all y_j. Such an analysis allows us to present the measurement result any of k indicators $X_1, \ldots, X_n$ as a linear combination with n influencing factors (factor Y and $(n-1)$ factors that are the remaining control indicators)

$$X_i = z_1\beta_{1i} + \ldots + z_n\beta_{ni} + \varepsilon_i, \tag{2.5}$$

where $\beta_{1i},\ldots\beta_{ni}$ are known constant coefficients, $i = \overline{1,N}$, and $z_1,\ldots z_n$ are variables that depend on the relevant influencing factors.

The type of analysis of variance of measurement results depends on the choice of coefficients β_{ji}, $j = \overline{1,n}$ and determines a specific mathematical model of the measurement object:

1. Model of multivariate cross-classification (parametric or variance component) – if β_{ji} can only take the value 0 or 1
2. Multiple regression model – if $\beta_{ji} \in [-\infty, \infty]$ for all $j = \overline{1,k}$ and $i = \overline{1,n}$
3. Covariance model – if some of the coefficients β_{ji} can take on the value only 0 or 1, and the rest – any values in the interval $[-\infty, \infty]$

Figure 2.1 shows a block diagram of the process of processing the results of N-fold measurements of vector components $\overline{X}$ for K states (e.g., levels of the control parameter Y) of the physical object of control.

Regarding the random residual ε_i of model (2.5), it should be noted that its variance σ_ε is a measure of the residual entropy of the X_i values after measurement (Ivanets et al. 2020) and the choice of the modelModels of the controlled value should be such that this variance is minimal.

2.2.4 Model of the Controlled Quantity at the Control Stage

The synthesis of the decision-making system for the control stage provides for the solution of the following tasks:

1. Choice of a model of discrimination of the results of single measurements of the components of a new vector $\overline{X}$ in order to assign this vector to one of the K clusters (or classes) $\overline{X}^{(j)}$, $j = \overline{1, K}$, that determine the level y_j of the control parameter

$$\forall \overline{X} \left[\overline{X} \in \overline{X}^{(j)} \rightarrow \overline{X} \in y_j \right]$$

2. Choice of the optimal number of clusters μ and the number of levels y_j for the maximum of expected information with fixed for all $j \in (2, K)$ residual variances σ_ε of model (2.5) $2 \leq \mu \leq K$
3. Assessment of the reliability of the results of discrimination

The discrimination model should have a certain robustness (insensitivity) to violation of a priori assumptions about the normality of the distribution laws of the vector components $\overline{X}$, and the model coefficients should be uniquely determined at the training stage, including the analysis of variance of the results of N-fold measurements. Since the maximum of the expected information about the level y_j of the parameter Y corresponds to the maximum of its likelihood function $L_j(\overline{X}^*)$, the choice of the cluster $\overline{X}^{(j)}$ is carried out from the condition

$$\overline{X}^* \in \overline{X}^{(j)}, \text{if } L_j(\overline{X}^*) = \sup \left[L(\overline{X}^*) \right], \quad (2.6)$$

where $L(\overline{X}^*) = \{ L_1(\overline{X}^*), L_\mu(\overline{X}^*) \}$.

Figure 2.2 shows a block diagram of processing the results of single measurements at the control stage.

Since $y_j^* \in \{ y_1, \ldots y_\mu \}$, formally, the absolute measurement error is determined by the expression (Shchapov and Avrunin 2011; Shchapov et al. 2015)

$$\Delta = y_j^* - y_j$$

and can be considered as a discrete random variable for which

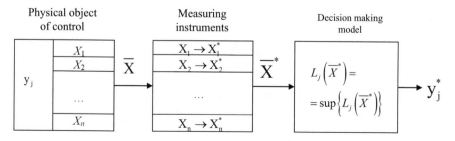

Fig. 2.2 Block diagram of processing the results of single measurements at the stage of measuring control. (Shchapov et al. 2015)

$$P(\Delta \neq 0) = P\left(y_j^* \neq y_j / y_j\right) = \alpha, \tag{2.7}$$

where α is the probability of a discrimination error of the first kind.

From (2.7) it follows that the measurement error Δ determines the reliability of the control (through the probability of error α). Thus, the use of procedures for making decisions about the level of a control parameter that is not amenable to direct, metrologically justified measurements can be considered as a method of indirect measurement. The implementation of this method can be carried out within the framework of computational training and discrimination procedures based on a multidimensional probabilistic model for a system of random variables $(Y, X_1, \ldots X_n)$.

2.3 Mathematical Model of a Dynamic Controlled Quantity with Discrete-Type Inhomogeneities

For such objects, the physical controlled quantities are functions of the observation time and are continuous random processes with varying degrees of nonstationarity due to the random nature of the second term in the framework of model (2.3). It is impossible to speak about the ergodicity of such a process, since the discrete-type inhomogeneity manifests itself in abrupt random changes in the mean values $\{X_v\}^{n-1}$ of the set of controlled quantities, stochastically related to the informative quantities X_l.

If the measured physical quantity is a random process $\xi(t)$, then, in order to obtain statistical conclusions about the characteristics of this process for fixed states π_1, $\ldots \pi_K$ of the control object, it is advisable to present it as a $\xi(t)$ truncated implementation on the observation interval T:

$$x(t) = \begin{cases} \xi(t), & t \leq T \\ 0, & t > T \end{cases}$$

This implementation can be replaced by the sum of quasi-deterministic random processes

$$x(t) = \sum_{i=1}^{n} \vartheta_i \varphi_i(t), \tag{2.8}$$

where $\vartheta_1, \ldots \vartheta_n$ is a set of unknown (random) parameters of the transformation model and $\varphi_1(t), \ldots \varphi_n(t)$ are the given deterministic functions.

Model (2.8) allows you to represent the process $\xi(t)$ in the space of parameters ϑ_i, $i = \overline{1, n}$, the numerical characteristics of which are functions of states $\pi_1, \ldots \pi_K$.

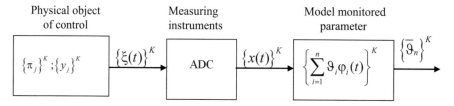

Fig. 2.3 Process flow block diagram $\{\xi(t)\}^k$ at the training stage. (Shchapov et al. 2015)

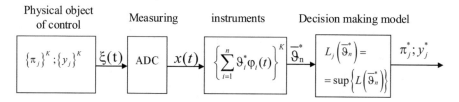

Fig. 2.4 Block diagram of processing processes $\xi(t)$ at the control stage. (Shchapov et al. 2015)

The parameters $\{\vartheta_i\}$ of the model are called the coordinates of a random process (Ivanets et al. 2019), which technically allow two types of transformations:

1. Discretization of the implementation $x(t)$ through correlation intervals τ_0 to obtain a sample $(x_1, \ldots x_N)$
2. Filtering $x(t)$ using a set of filters whose impulse transient characteristics are matched with the correlation function of the process $\xi(t)$ (to obtain independent signals at the filter outputs)

In any case, model (2.8) transforms the problem of monitoring any of the K states into discrimination of the vector of model parameters $\overline{\vartheta}_n = (\vartheta_1, \ldots \vartheta_n)$. If the states $\pi_1, \ldots \pi_K$ correspond to the levels $y_1, \ldots y_K$ of the physical parameter Y of the control object, then at the training stage, it is necessary to estimate the minimum number of samples $N_{\min}$, which ensures the given control reliability. At the control stage, as well as at the training stage, it is possible to use, to normalize the vector $\overline{\vartheta}_n$, normalizing transformations of the model parameters $\{\vartheta_i\}$ in statistics $\{S_i\}$, the mean values $\{\overline{S}_i\}$, $i = \overline{1, n}$ of which would be functions of the levels $\{y_j\}$, and the variances $\{\sigma_{si}^2\}$ would not depend on the state number $j = \overline{1, K}$.

Figures 2.3 and 2.4 show the block diagrams of processing, respectively, multidimensional (by states) processes $\{\xi(t)\}^k$, the learning stage, and one-dimensional (by the j-th state) process $\xi(t)$ – the control stage. The measuring instrument is an analog-to-digital converter (ADC). Normalizing conversions are omitted.

For the control stage, the condition must be satisfied:

$$\forall \overline{\vartheta}_n \left[\overline{\vartheta}_n \in \overline{\vartheta}_n^{(j)} \to \overline{\vartheta}_n \in y_j \right] \tag{2.9}$$

2.4 Mathematical Model of the Controlled Quantity in Conditions of Continuous and Discrete Types of Inhomogeneities

Such a value is represented by a sequence of measurement results ordered in time, in relation to which the following can be said:

- The component $F_l[\cdot]$ of model (2.4) introduces a priori uncertainty in the form of a trend (long-term influence) (Shchapov et al. 2018), reflecting gradual parametric changes in the physical properties of the control object.
- The component ε_l determines short-term, more or less regular, fluctuations in the measurement results relative to the trend due to the unavoidable factorial influence of stochastic relationships between the parameter Y and the set of physical properties of the control object in the form of a subset $\{X_l\}^{n-1}$.
- Unlike a dynamic object with discrete heterogeneity, this model is characterized by a very long observation time and corresponds to objects of preventive control (controlled operation) (Shchapov 2005).

The most adequate model of the measurement object under consideration is the time series model (Shchapov and Avrunin 2011). In terms of analyzing the uncertainty of model (2.4), additional difficulties may arise if measurements are made at unequal intervals, which is most often the case in practice.

2.4.1 Dynamic Controlled Variable Model at the Training Stage

If the random remainder of model (2.7) is stationary $\varepsilon_l(t)$, the main probabilistic properties of this model are determined by the terms $F_l[\cdot]$ and ε_l, and the general model (2.7) is transformed into model (2.10). At the training stage, it is necessary to remember that in expansion (2.10) the first and second terms are not independent but are stochastically related through the control parameter Y. This means that the additivity of model (2.10) is provided at the level of only average values, which allows presetting (based on the properties of a physical object) the type of function $F_l[\cdot]$. A variance analysis of the functional influence of time t on the function $F_l[\cdot]$ is possible only within the framework of all three terms of model (2.4). Figure 2.5 is a block diagram of the learning process.

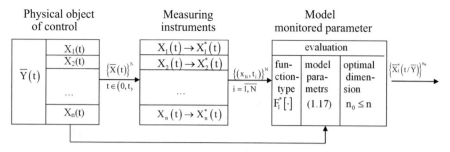

Fig. 2.5 Structural diagram of the analysis of probabilistic properties of controlled value at the stage of training the control system. (Shchapov et al. 2015)

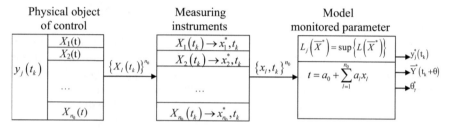

Fig. 2.6 Block diagram of processing the results of single measurements at the stage of control and technical diagnostics. (Shchapov et al. 2015)

2.4.2 Model of a Dynamic Object of Measurement at the Control Stage

The control stage provides, formally, the solution of three tasks for a point in time $t_k > t_N$:

1. Assessment of the real functional state of the control object in the form of a quantity $Y(t_k)$
2. Assessment of the real functional state of the control object in the form of a quantity $Y^*(t_k + \theta)$ for time $t_{N+1} = t_N + \theta$;
3. Forecasting the residual resource θ_γ that ensures the normal functional state of the object with a confidence level γ (Shchapov and Avrunin 2011) at norm Y_H, $(P[Y^*(t_k + \theta_\gamma) \in Y_H] = \gamma)$

Figure 2.6 is a block diagram of the control process that ensures the solution of the listed tasks.

2.5 Practical Application of Mathematical Models of Dynamic Measurement Objects

As an example of using these mathematical approaches, the results of experimental studies of transformer oils were processed. The practical use of mathematical models of analysis of variance, especially a model with two-dimensional observations, was aimed at reducing the effect of mixed-type statistical heterogeneity (model (2.4)) on the results of measuring operational monitoring of the state of liquid insulation (transformer oil) of high-voltage operating power equipment. Optimal training and control procedures (to the maximum of the expected information) made it possible to solve the problems of not only increasing the reliability of control but also developing recommendations for reorganizing the system for preventive testing of oil-filled high-voltage power equipment, in terms of the frequency of operational control of transformer oils.

Ensuring the reliability of electric power facilities during their long-term operation is a state task. Such objects belong to the class of repairable, for which the technical resource and durability of functioning directly depend on the efficiency of the maintenance and repair system. In this case, the issues of ensuring the maximum reliability of decisions about the type of the current technical state of the controlled object are of particular importance. Such solutions can be used both for planning preventive work and for predicting possible equipment malfunctions.

However, any control presupposes the availability of a priori information about the values of the input values of the control system for various technical states of the equipment. Such monitoring should be aimed, first of all, at identifying both gradual and sudden failures caused by equipment wear. Common to all control objects in the electric power industry is a large specific weight of insulation, including liquid insulation, for which aging processes are a factor that degrades the reliability of equipment.

Existing physical models of insulation aging processes, as a rule, take into account the change in the chemical structure of the dielectric under the influence of factors such as operating temperature and electric field strength. At the same time, the characteristics of insulation, especially liquid insulation, are quite complex and imply both deterioration and improvement of its certain physical properties. To train control systems, only using physical models means introducing systematic errors into decision-making procedures, which will necessarily reduce the reliability of control. This is all the more ineffective, since the insulation properties of the equipment are influenced not only by the operating time but also by such physical factors as temperature and changes in electrical load modes. In this case, the problems of the operability state are of particular importance, which create difficulties in assessing the residual life of the equipment.

The interpretation of the results of any measurements in the course of preventive maintenance and control of power equipment depends on the following a priori specified conditions for obtaining primary measurement information:

1. The input quantities used should be as sensitive as possible to the influence of not only the operating time but also physical, periodically (or constantly), acting factors that change the properties of the equipment.
2. Mathematical models of the change in the average values of the input quantities under the influence of the operating time must be justified not only physically but also statistically.
3. The input values should be divided into subsets that provide the maximum amount of expected measurement information for the given classes of technical condition of the equipment.
4. The used subset of input quantities must be robust (insensitive) to a certain inadequacy of physical and mathematical models of parametric changes caused by long-term operation of equipment.
5. Probabilistic models of changes in the values of controlled quantities should be known by the types of equipment and operating conditions, at least with an accuracy up to the values of the parameters of these models.

The purpose of measuring technical control in the power industry is to prevent the development of emergency situations based on the use of not only more advanced decision-making procedures but also more informative input values of the control system with a priori studied probabilistic properties for specific types of insulation.

A feature of the research is the practical use of parametric and random models of factor influence on controlled quantities in conditions of information uncertainty of multidimensional arrays of measurement information. At the same time, statistically reliable conclusions were obtained on the probabilistic properties of subsets of the input quantities, which make it possible to provide the maximum amount of expected measurement information on the qualitative states of liquid insulation.

The results of periodic measurements of physical input quantities reflecting the properties of transformer oil can be considered as time series containing information about the change in the value of any of the physical quantities, from the moment of filling the oil to the current moment (Shchapov and Chunikhina 2011). Moreover, the most interesting will be the series containing the maximum number of measurement results.

From the point of view of the physical adequacy of the processes, operational data have a significant advantage over the data obtained experimentally in laboratories, since they most fully reflect the impact of real operational factors.

As the initial sample, the results of preventive tests of transformer oil in four different areas were analyzed. In total, the analysis of the results of periodic tests of transformer oil for 91 transformers with voltage on the high side of 110 kV, rated power from 6.3 to 40.5 mV•A, types TRDN, TDTN, TDNG, TND, TDTNG, TDN, and TMN with various stresses, on the middle and low sides. The distribution of the number of transformers by region, capacity, type, and rated voltage is shown in Table 2.1.

Each studied oil chart contains a set of test dates and the corresponding results of measurement of quality indicators (input values of the control system). A

Table 2.1 Distribution of information on oil-filled transformers

Region	Rated power of transformers, mV•A	Transformer type	Rated voltage of transformers, kV	Number of transformers
1	40,5	TDTN	110/35/6	1
	40	TDTN	110/35/6	1
	25	TDTN	110/35/6	2
			110/35/10	4
2	6,3	TMN	110/10	1
			110/6	1
	20	TDNG	110/6	1
	25	TDTN	110/35/6	5
			110/35/10	4
	40	TDTN	110/35/10	1
3	6,3	TMN	110/10	1
	20	TDTN	110/35/10	4
		TDTNG	110/35/10	1
	25	TDTN	110/35/6	2
			110/35/10	
	40	TRDN	110/6	1
		TDTN	110/35/10	4
4	40,5	TDTN	110/35/6	2
	40	TND	110/10	1
		TRDN	110/6	5
			110/10	3
		TDNG	110/35/10	1
		TDTN	110/35/6	8
	32	F	110/6	1
		TDNG	110/6	1
		TRDN	110/10	2
			110/6	3
	25	TDNG	110/6	1
		TRDN	110/10	3
			110/6	5
		TDTN	110/35/10	3
			110/35/6	8
			110/10/6	5
	20	TDNG	110/6	1
	16	TDN	110/10	2
Total				91

two-dimensional array of measurement dates and quality index values is a time series.

Since the investigated transformers were put into operation at different time periods, when the rows of indicators are combined into a common array, the rows shift along the ordinate axis, which introduces certain inconveniences for further data processing.

2 Increasing the Reliability of Diagnosis and Control in the Uncertainty...

Table 2.2 Distribution of transformers by service life

Number of transformers	Service life of transformers			
	Up to 10 years	10–20 years	20–30 years	More than 30 years
	11	49	21	10

Therefore, when forming the time series, the test dates (date, month, year) were converted into the life of the transformer, while the first value (commissioning of the transformer) was equal to zero, and all subsequent values were determined by summing the number of days between the test dates divided by the number of days in a year.

The number of days between tests was calculated taking into account the number of calendar days in months and in the corresponding year (365 or 366 days). The maximum service life of the investigated transformers was 39.419 years, and the minimum service life was 5.95 years. The distribution of transformers by service life is given in Table 2.1. Table 2.2 shows that the possible statistical uncertainty of the controlled quantities includes a discrete component.

The list of controlled quantities as standard indicators and the volume of sample values for each indicator is given in Table 2.3. From Table 2.3 it follows that the sample is heterogeneous both in the number of rows and in the total values of the number of feature values. This heterogeneity is due to several factors. The control of the X_4, X_5, and X_6 values was not strictly regulated. The X_7 indicator was monitored in regions "1" and "3" as an additional indicator. The insignificant volumes of sample values of indicators X_8, X_9, X_{10}, and X_{11} can be explained by the fact that these indicators have two possible implementations, the absence of this indicator in the oil or its presence. As a rule, the appearance of these indicators indicates a deep aging of the oil.

With regard to the results of the chemical analysis of soluble gases (signs X_{12}–X_{19}), the following features should be noted: indicators X_{12} and X_{13} (content of CO and CO_2) are present in all transformers, over the entire period of operation. As for the gases of the hydrocarbon series and hydrogen, it should be noted that these gases were not found in all investigated transformers. Table 2.3 also presents the values of estimates r^* of the paired correlation coefficients of control indicators $X_1,...X_{19}$ with the time of observation of the T_H (Ivanets and Morozova 2021).

From Table 2.3 it can be seen that the total sample size for transformers gives 460 subsets of the initial data, with a total volume of 6267 values representing the dependence of the sample values of indicators on the operating time.

Presented in Table 2.3 input values (hereinafter indicators, which correspond to a number from 1 to 19: $X_1,...X_{19}$) are, in fact, estimates in the form of time series, nonstationary, continuous, random processes with discrete time and can be used as training samples in the development of procedures for measuring control of technical condition and predicting parametric violations. The latter are understood as the discrepancy between the measured values of the indicators $X_1,...X_{19}$ permissible standard values, and one of the boundaries of the tolerance zone (lower or upper, depending on the properties of the indicator) is the maximum permissible (Shchapov et al. 2015).

Table 2.3 List of quality indicators of transformer oil and values

| Input value | Oil quality indicator | T_H, years | Sample size | | r^* |
			Number of rows	Number of values	
1	Flash point	2,09	89	971	- 0,067
2	Acid number	18,10	90	969	0,503
3	Breakdown voltage	2,04	90	950	- 0,066
4	tgδ at 20 °C	0,92	11	66	0,114
5	tgδ at 70 °C	5,35	23	157	0,395
6	tgδ at 90 °C	3,18	12	108	0,295
7	Oil color	11,36	16	194	0,634
8	Content of water-soluble acids	0,94	7	83	0,104
9	Content of mechanical impurities	1,84	1	9	0,57
10	Sludge content	2,30	1	5	0,799
11	Oil moisture content	1,30	10	93	- 0,135
12	CO content	4,97	8	153	0,375
13	CO_2 content	3,49	26	721	0,129
14	CH_4 content	2,15	11	247	- 0,136
15	C_2H_2 content	2,52	14	343	- 0,135
16	C_2H_4 content	1,59	21	559	- 0,067
17	C_2H_6 content	0,38	11	211	0,026
18	The sum of hydrocarbon gases	2,96	16	393	- 0,148
19	H_2 content	0,36	3	35	- 0,063

Adapted from Bondarenko et al. (2007)

The statistical heterogeneity of the source data can be caused by many reasons:

(a) Deviations of physical aging models from linear dependences of the quality index (e.g., saddle-like dependence of the dielectric loss tangent on the aging time, an increase in the flash point value at the beginning of operation due to evaporation of light fractions, an increase in breakdown voltage after oil filling due to moisture escape, etc.)
(b) The influence of external influences (preventive measures to improve the properties of the oil during operation, lightning or switching overvoltages, phase imbalances, short circuits in an electrically connected network, overloads of transformers, etc.)
(c) Metrological failures of measuring equipment (decrease in accuracy class)

(d) Methodological errors and slips during measurements
(e) Errors in recording measurement results

The performed analysis of oil maps for all investigated transformers made it possible to establish that about 40% of all distortions are due to external influences on the liquid insulation during operation and about 30% of distortions appeared as the result of sampling oil from various points of the transformer. For the remaining 30% of the distortions, it was not possible to reliably establish the cause due to the lack of records in the maps.

External influences should be understood primarily as preventive measures to improve the condition of the oil during operation (drying, regeneration, oil refilling, replacing silica gel, or changing oil), that is, artificial intervention of substation personnel in the aging process in order to reduce the rate of its course. Such interference is especially characteristic of transformer oil quality indicators, such as acid number, moisture content, breakdown voltage, and dielectric loss tangent.

At the same time, for gases dissolved in oil, in particular, it concerns gases of the hydrocarbon series; the presence of distortions can mainly be explained as a consequence of the processes of gas evolution and recombination under the influence of operating factors, i.e., the reaction of the insulating medium to external disturbances. As for the distortions caused by different sampling points of the oil samples, it is quite obvious that the oil in the tank is not a homogeneous medium, and its properties are uneven depending on the sampling location. It is clear that from the point of view of operational reliability, the change of sampling sites is absolutely necessary, despite the fact that the results obtained in this case introduce a noticeable noise into the original sample. The mechanism of the appearance of distortions is clearly illustrated in Fig. 2.7 using the example of the indicator X_2 of measurement control.

In Fig. 2.7, three characteristic areas can be clearly distinguished. The first area, from the moment of oil filling up to 20 years of operation, is characterized by a

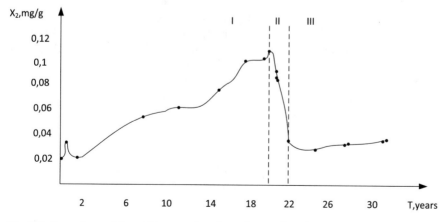

Fig. 2.7 Dependence of the acid-base number of transformer oil on the operating time of one of the "four regions" transformers. (Adapted from Bondarenko et al. 2007)

steady increase in the acid number value due to the aging and oxidation processes of the oil. In the second section from 20 to 22 years, there is a sharp decrease in the value of this indicator. The reason for the decrease in the value of the acid number was the drying and replacement of the silica gel (21–22 years). Then there was an oil change (for 23 years). And then, in the third section, the aging of the replaced oil occurs. Similar measures lead to distortion of the sign of the pair correlation coefficient.

To reduce the a priori uncertainty of the results of operational measurements carried out during the normative preventive tests of transformer oils, the processing of the initial data arrays at the stage of training the control system was carried out according to the following algorithm (Shchapov et al. 2015):

1. Elimination of time series, distorted by the sign of the pair linear correlation coefficient (series were eliminated, the regression models of which did not correspond to the physical models for measuring mean values during long-term operation) (Shchapov et al. 2018)
2. Checking the remaining time series for homogeneity across groups of equipment (a statistical homogeneity check model was used, taking into account parametric gradual equipment failures)
3. Formation of the optimal, according to the maximum of the objective function β_p, a system of indicators of measuring control with ranking of indicators in descending order of the amount of expected measuring information I about parametric changes during operation (Shchapov 2005)
4. Assessment of additive, multiplicative, and full factorial influences on the results of measuring the values of control indicators based on covariance models
5. Evaluation of absolute and reduced (to the range of sample values of indicators) measurement errors, taking into account the additive and multiplicative factor influences
6. Forecasting the moments of parametric violations at a given reliability

Table 2.4 shows the numerical characteristics of the results of two-dimensional measurements and the parameters of partial regressions for three physicochemical (X_1, X_2, X_3) and two indicators of chromatographic analysis (X_{12}, X_{13}). Table 2.4 shows the results of the covariance analysis of the regression models for all 19 measures of measurement control (Shchapov et al. 2018).

In these tables, all F-statistics are based on mathematical variance-decomposition models. Statistics F_{WG}, F_W, F_G, and F_α are defined by expressions

$$F_{WG} = \frac{S_{WG}}{S_R}(N - 2k) \tag{2.9}$$

$$F_G = \frac{S_G}{S_R}\left(\frac{N - 2k}{k - 2}\right); \tag{2.10}$$

Table 2.4 Results of covariance analysis of linear models of transformer oil quality indicators, data array M_a (initial data without preliminary processing)

Oil quality indicators	k	N	Degrees of freedom values							F-statistic values							I, bit
			V_0	$V_{\omega\sigma}$	V_σ	V_ω	V_r	V_α	$V_{r\alpha}$	F_0	$F_{\omega\sigma}$	F_σ	F_ω	F_α	F_{Σ^0}	
1	89	971	1	1	87	88	793	88	881	19,205	72,878	30,275	7,213	18,979	18,986	2,1709
2	90	969	1	1	88	89	789	89	878	953,92	419,63	13,568	4,554	13,329	11,342	4,9457
3	90	950	1	1	88	89	770	89	859	6,327	25,823	4,705	2,288	4,36	3,616	1,4277
4	11	66	1	1	9	10	44	10	54	3,13	22,437	12,842	5,565	7,478	9,683	1,0231
5	23	157	1	1	21	22	111	22	133	78,109	33	7,237	5,747	4,17	7,0781	3,1544
6	12	108	1	1	10	11	84	11	95	30,817	92,494	6,045	7,678	7,841	10,791	2,4958
7	16	194	1	1	14	15	162	15	177	461,91	0,0022	28,038	8,836	15,725	17,502	4,4383
8	7	83	1	1	5	6	69	6	75	1,966	10,163	4,946	12,262	3,0594	9,039	0,9723
11	10	93	1	1	8	9	73	9	82	1,844	4,201	1,345	1,128	1,639	1,396	0,7332
12	8	153	1	1	6	7	137	7	144	25,383	0,878	1,835	0,884	1,708	1,291	2,3607
13	26	721	1	1	24	25	669	25	694	16,385	30,928	6,333	4,34	6,531	5,829	2,0477
14	11	247	1	1	9	10	225	10	235	6,841	4,355	10,031	4,22	8,322	6,842	1,4856
15	14	343	1	1	12	13	315	13	328	7,295	0,735	3,567	2,323	3,182	2,836	1,525
16	21	559	1	1	19	20	517	20	537	2,966	2,006	4,262	2,008	3,999	3,0792	0,9951
17	11	211	1	1	9	10	189	10	199	0,276	10,365	15,4459	6,306	11,802	10,628	0,1766
18	16	393	1	1	14	15	361	15	376	10,896	1,393	4,941	3,627	4,258	4,166	1,7862
19	3	35	1	1	1	2	29	2	31	0,123	0,864	0,0243	0,499	0,459	0,472	0,0838

$$F_{\mathrm{W}} = \frac{S_{\mathrm{W}}}{S_{\mathrm{R}}} \left(\frac{N - 2k}{k - 2} \right); \tag{2.11}$$

$$F_\alpha = \left(\frac{S_{\mathrm{WG}} + S_{\mathrm{G}}}{S_{\mathrm{R}} + S_{\mathrm{W}}} \right) \left(\frac{N - k - 1}{k - 1} \right). \tag{2.12}$$

We used the decomposition of the sum S of the squared deviations of Y_{ji} observations from the total mean $\overline{Y}$ into five terms:

$$S = S_0 + S_{\mathrm{WG}} + S_{\mathrm{G}} + S_{\mathrm{W}} + S_{\mathrm{R}},$$

where

$S_0 = w_0 B_0^2;$
$S_{\mathrm{WG}} = \frac{w_c w_m}{w_0} (B_c - B_m)^2;$
$S_{\mathrm{G}} = \sum_j n_j \left[\overline{Y}_j - \overline{Y} - B_m \left(\overline{t}_j - \overline{t} \right) \right]^2;$
$S_{\mathrm{W}} = \sum_j w_j (B_j - B_c)^2;$
$S_{\mathrm{R}} = \sum_j \sum_i \left[\overline{Y}_{ji} - \overline{Y}_j - B_j \left(t_{ji} - \overline{t} \right) \right]^2.$

The constants $w_0, w_m,$ and w_c are related by the equation

$$w_0 = w_m + w_c,$$

where

$w_0 = \sum_j \sum_i \left(t_{ji} - \overline{t} \right)^2,$
$w_m = \sum_j n_j \left(\overline{t}_j - \overline{t} \right)^2,$
$w_c = \sum_j \sum_i \left(t_{ji} - \overline{t}_j \right)^2,$

t_{ji} is the countdown time of the Y_{ji} value

The angular coefficients B_0, B_m, and B_c are determined by the expressions

$$Bc = E[Bj];$$

B_0 is the angular coefficient of the regression plotted over the entire total time series;

$$B_m = w_m^{-1} (w_0 B_0 - w_c B_c).$$

For each quality indicator, informative F-statistics were calculated, including for the additive F_α and multiplicative F_{W} factor influence on the measurement results.

Table 2.4 shows the values of the F-statistics obtained in the experiment. In the same table, the values of all test F-statistics are given.

From Table 2.4 it can be seen that the maximum information about technological violations is carried by independent dimensionless statistics F_{WG}, F_G, and F_W, and statistics F_0 and $F_{0\Sigma}$ (reflect the general factorial influence, including metrological uncertainty) have the form

$$F_0 = w_0 B_0^2;$$ (2.13)

$$F_{0\Sigma} = \left(\frac{S_{WG} + S_G + S_W}{S_R} \right) \left(\frac{N - 2k}{2k - 2} \right).$$ (2.14)

Table 2.5 shows the results of the covariance analysis of the data set, from which the time series are removed, the regression models of which have statistically insignificant slope coefficients, which correspond to the training model without taking into account a priori uncertainty. Tables 2.4, 2.5 and 2.6 represent (the last column) estimates of the amount I of expected measurement information on parametric changes for each control indicator (in bits). From these tables, it is clearly seen that the elimination of misses (by the distorted coefficient of pair correlation) and the check for stochastic homogeneity provide the maximum values of the expected measurement information (Table 2.6). Covariance analysis shows, the same, that the additive factor influence is higher than the multiplicative influence ($F_\alpha > F_W$). This gives grounds to assert that the nature of the uncertainty of parametric changes is, in most cases, an abrupt character and is determined by the initial conditions of the liquid insulation operation.

Table 2.7 shows the absolute (ΔT) and reduced (γ_T) errors in predicting the moments of time of parametric violations for subsets of indicators (input values of the control system) included in the multiple linear regression equation. The errors are calculated for the confidence level $P = 0.95$ and the standard service life $T = 25$ years.

2.6 Conclusion

The model for collecting and processing information in the system of operational monitoring of the states of liquid insulation of operating high-voltage power equipment has been improved, which makes it possible to increase the reliability of monitoring of parametric violations of the quality of transformer oils. The effectiveness of the application of methods of statistical estimation of the minimum permissible number of input quantities in the control and forecasting system of parametric violations, while ensuring the maximum of expected information at a given control reliability, has been proved.

Table 2.5 The results of the covariance analysis of linear models of the quality indicators of transformer oil, the M_6 data array (obtained from the initial M_a data array by screening out time series, distorted by the sign of the pair correlation coefficient)

Oil quality indicators	k	N	Degrees of freedom values							F-statistic values						I, bit
			V_0	$V_{\omega\sigma}$	I, bit	V_ω	V_r	V_α	$V_{r\alpha}$	F_0	$F_{\omega\sigma}$	F_σ	F_ω	F_α	F_{Σ^0}	
1	56	614	1	1	54	55	502	55	557	180,75	210,57	41,958	4,413	33,674	24,718	3,7668
2	77	794	1	1	75	76	640	76	716	1247,5	304,52	12,203	2,326	14,069	9,188	5,1401
3	56	501	1	1	54	55	389	55	444	148,88	0,3138	7,414	2,143	6,381	4,714	3,6138
4	9	49	1	1	7	8	31	8	39	11,106	13,22	10,474	3,761	6,905	7,289	1,7988
5	17	113	1	1	15	16	79	16	95	105,08	34,828	4,756	4,539	4,157	5,587	3,3662
6	12	108	1	1	10	11	84	11	95	30,817	92,494	6,0457	7,678	7,841	10,791	2,4958
7	15	185	1	1	13	14	155	14	169	441,39	0,1311	25,901	8,742	14,659	16,401	4,5412
8	7	83	1	1	5	6	69	6	75	1,966	10,163	4,946	12,262	3,0594	9,039	0,9723
11	6	41	1	1	4	5	29	5	34	2,42 10-6	7,47	1,371	0,43	2,827	1,51	0,7895
12	7	125	1	1	5	6	111	6	117	26,803	2,92	2,009	0,283	2,135	1,17	2,3986
13	19	541	1	1	17	18	503	187	521	22,902	57,455	4,89	1,444	7,692	4,627	2,2895
14	5	68	1	1	3	4	58	4	62	0,972	9,527	25,938	0,034	14,381	15,435	0,49
15	9	190	1	1	7	8	172	8	180	0,05983	1,617	2,661	0,0227	2,645	1,276	0,9982
16	10	211	1	1	8	9	191	9	200	7,55	3,885	9,416	2,583	8,216	5,692	1,548
17	9	168	1	1	7	8	150	8	158	0,0004	13,407	14,929	6,189	11,672	10,464	0,0002
18	9	161	1	1	7	8	143	8	151	0,7803	9,512	8,148	2,186	7,826	5,252	0,416
19	3	35	1	1	1	2	29	2	31	0,123	0,864	0,0243	0,499	0,459	0,472	0,0838

2 Increasing the Reliability of Diagnosis and Control in the Uncertainty...

35

Table 2.6 Results of covariance analysis of transformer oil quality models, M_o data array (obtained from an M_6 array by extracting statically homogeneous time series)

Oil quality indicators	k	N	Degrees of freedom values							F-statistic values						I, bit
			V_0	$V_{\omega\sigma}$	V_σ	V_ω	V_r	V_α	$V_{r\alpha}$	F_0	$F_{\omega\sigma}$	F_σ	F_ω	F_α	F_{Σ^0}	
1	12	160	1	1	10	11	136	11	147	246,156	55,521	7,041	5,491	8,568	8,47	4,2288
2	29	289	1	1	27	28	231	28	259	3275,43	59,057	9,856	5,824	7,632	8,719	6,3819
3	27	257	1	1	25	26	203	26	229	503,127	42,264	24,626	6,361	15,73	15,833	4,4888
4	3	20	1	1	1	2	14	2	16	39,4	3,0401	8,337	4,321	4,0199	5,0051	2,6681
5	3	30	1	1	1	2	24	2	26	70,9117	0,31329	2,5261	1,5904	1,358	1,505	5,676
6	6	61	1	1	4	5	49	5	54	69,979	5,755	1,775	1,171	2,531	1,871	3,3224
7	10	127	1	1	8	9	107	9	116	487,647	4,866	10,746	12,079	5,427	11,086	4,8276
8	3	44	1	1	1	2	38	2	40	45,984	0,473	0,466	0,506	0,481	0,488	2,777
12	3	74	1	1	1	2	68	2	70	20,342	1,107	0,237	0,0354	0,691	0,353	2,2078
13	6	182	1	1	4	5	170	5	175	38,426	2,11	4,874	0,502	4,384	2,412	2,6505

Table 2.7 Prediction errors of parametric violations of the quality of transformer oil

Indicator		Subsets of indicators			
Type	I (bit)	Regressors	β_p	ΔT, years	γ_T, %
X_2	6,3819	X_2	–	6,20	24,80
X_5	5,676	X_2, X_5	4074,82	4,98	19,92
X_3	4,4888	X_2, X_5, X_3	4822,43	3,40	13,60
X_1	4,2288	X_2, X_5, X_3, X_1	5557,86	2,52	10,08
X_8	2,777	X_2, X_5, X_3, X_1, X_8	5787,53	2,43	9,72
X_{13}	2,6505	X_2, X_5, X_3, X_1, X_8, X_{13}	4894,22	2,46	9,84
X_{12}	2,2078	X_2, X_5, X_3, X_1, X_8, X_{13}, X_{12}	1410,71	2,53	10,12

References

Bondarenko VE, Shchapov P, Shutenko OV (2007) Improving the efficiency of operational measuring control of transformer oils. NTU "KhPI", Kharkiv

Ivanets OB, Morozova IV (2021) Features of evaluation of complex objects with stochastic parameters. In: 11th international conference on advanced computer information technologies, ACIT 2021. 15–17 September 2021. IEEE, pp 159–162

Ivanets O, Morozova IV, Tereshchenko YM (2019) Approach to assessing quality indicators. In: International conference on advanced optoelectronics and lasers CAOL*2019. 06–08 September 2019. IEEE, pp 666–670

Ivanets OB, Shchapov PF, Sevryukova OS (2020) Dynamic properties of the time series of biomedical measurement. Sci Based Technol 46(2):236–244. https://doi.org/10.18372/2310-5461.46.14811

Shchapov PF (2005) Optimization of the space of information parameters based on the analysis of variance covariance models. Electr Eng Electromech 2:59–62

Shchapov PF, Avrunin OG (2011) Increase in the reliability of control and diagnostics of objects in conditions of uncertainty. KhNADU, Kharkiv

Shchapov PF, Chunikhina TV (2011) Minimization of the time of preventive control of the parameters of liquid insulation of power facilities. East-Eur J Adv Technol 4(8(52)):58–60

Shchapov PF, Migushchenko RP, Kropachek OY (2015) Theoretical and practical ambush of control systems and diagnostics of folding industrial objects. NTU "KhPI", Kharkiv

Shchapov PF, Migushchenko RP, Kropachek OY et al (2018) Investigation of correlation models of spectral nonstationarity of random signals. Metrol Instrum 5(73):11–14

Chapter 3
Environmental Impact Assessment of the Planned Activity of Aviation Transport

Viktoriia Khrutba, Tetiana Morozova, Anna Kharchenko, Inesa Rutkovska, and Alla Herasymenko

Nomenclature

EIA   Environmental impact assessment
PCIs   Projects of Common Interest
ECEIA   Electronic calculator environmental impact assessment
FFS   Fuel filing station
ERS   Emergency rescue station

3.1 Introduction

Environmental impact assessment (EIA) is a form of environmental assessment, an internationally recognized tool for preventive environmental policy, aimed at studying, analyzing, and evaluating planned activities to ensure environmentally sustainable development in accordance with international and domestic environmental legislation (Pro 1991).

Thus, the environmental impact assessment of aviation infrastructure requires systematic research and justification of the need for environmentally oriented management of design and implementation of works on construction, reconstruction, and operation of aviation infrastructure, in order to create conditions for improving their efficiency while minimizing negative consequences of anthropogenic intervention in the ecosystem.

V. Khrutba · T. Morozova · A. Kharchenko (✉) · I. Rutkovska · A. Herasymenko
National Transport University, Kyiv, Ukraine

© The Author(s), under exclusive license to Springer Nature Switzerland AG 2023
S. Boichenko et al. (eds.), *Sustainable Transport and Environmental Safety in Aviation*, Sustainable Aviation, https://doi.org/10.1007/978-3-031-34350-6_3

The objective of the paper is to develop and implement mechanisms for the implementation of environmental impact assessment for transport facilities and its implementation for projects of planned air transport activities.

To achieve this goal, the following tasks were formulated:

- Analyze domestic and international legislation on environmental impact assessment.
- Define assessment criteria and indicators of environmental impacts for air transport facilities.
- Develop an electronic service for calculating environmental impact assessment criteria for air transport facilities.
- Assess the impact on individual elements of the environment of the Flight Zone No. 2 of SE IA "Boryspil" Reconstruction Project.

The object of research is the processes of environmental impact of the planned activities of transport facilities.

The subject of research is air transport facilities construction and reconstruction projects.

3.2 Legislation on Environmental Impact Assessment

To meet international requirements for identifying the nature, intensity, and degree of danger of environmental impact, Ukraine has to adapt its national legislation to EU standards in accordance with the appropriate Association Agreement with the EU.

Environmental assessment in international law is based on the following:

1. The general principles of international law, which were implemented in the cases of Trail Smelter (1941), Nagymaros-Gabcikovo (1997), Pulp Mill (2010), Nicaragua vs. Costa Rica (2015). In the Trail Smelter case, e.g., the arbitration award required a steel plant operator to reimburse losses if emissions exceeded preset limits, regardless of the damage they could cause.
2. On the principles of the Rio Declaration on Environment and Development, namely, integration (principle 4), environmental assessment (principle 17), liability for transboundary environmental damage (principle 2), transboundary procedure (principles 18 and 19).

The above principles form the basis of documents on the use of natural resources to harmonize national and cross-border procedures. These are, e.g., the EIA Directive (1985), the Convention on Transboundary Effects (Espoo) (1991), the SEA Directive (2001), and the SEA Protocol (2003).

European environmental legislation in terms of environmental impact assessment is developing dynamically. The first EIA Directive in 1985 was amended three times and replaced by the 2011 Directive (Directive 2011/92/EU of the European Parliament and of the Council of 13 December 2011 on the assessment of the effects of certain public and private projects on the environment), which was also amended in 2014 (Directive 2014/52/EU of the European Parliament and of the Council of 16 April 2014).

In European legislation, the procedure for environmental impact assessment (Environmental Impact Assessment of Projects. Guidance on Screening) provides for requirements for authorities to provide information on the expected environmental impact (scoping stage); the obligation of the customer to provide information on the expected environmental impact of the planned economic activity, i.e., the EIA report (EIA report – Annex IV); informing and consulting the competent authorities and the public (and the recipient country); and submission by the authorized authorities of a permit document on the commencement of the planned activity, taking into account the results of the consultations. Notices of the permitting decision are made public, after which the public can appeal the decision in court.

There are a number of guidelines and common European standards for the effective implementation of the requirements of the Directive, namely:

- Guidelines for determining the need for an environmental impact assessment (EIA) during the activity (Environmental Impact Assessment of Projects. Guidance on Screening, Clarification of the application of Article 2(3) of the EIA Directive, EIA – Guidance on Screening)
- Guideline for determining the scope and detail of EIA (Screening checklist EIA – Guidance on Scoping – 2001)
- Guideline for preparation of the EIA report (Environmental Impact Assessment of Projects. Guidance on Scoping, Environmental Impact Assessment of Projects. Guidance on the preparation of the Environmental Impact Assessment Report)
- Documents for determining the cumulative impact (EIA Review Check List – 2001)
- Documents for determining the transboundary impact (Streamlining environmental assessment procedures for energy infrastructure Projects of Common Interest (PCIs), Guidelines on the Assessment of Indirect and Cumulative Impacts as well as Impact interactions, Communication from the Commission – Trans-European networks: Towards an integrated approach)
- Clarification on certain types of projects (Commission guidance document on streamlining environmental assessments conducted under Article 2(3) of the EIA Directive, Guidance on Integrating Climate Change and Biodiversity into Environmental Impact Assessment, Interpretation of definitions of project categories of annex I and II of the EIA Directive, Interpretation of definitions of certain project categories of annex I and II of the EIA Directive, Interpretation suggested by the Commission as regards the application of the EIA Directive to ancillary/ associated works, Application of the EIA Directive to projects related to the exploration and exploitation of unconventional hydrocarbon, Application of EIA Directive to the rehabilitation of landfills)

- Climate and biodiversity impact assessments (Guidance on the Application of the Environmental Impact Assessment Procedure for Large-scale Transboundary Projects)
- Impact assessments on landscapes and other materials

The Environmental Standards Directive in the water sector, the Air Quality Directive 2008/50/EC, and other European Commission documents are a well-developed system of standards aimed at achieving good quality of all environmental components and explaining potential environmental threats from economic activities and ensuring inexhaustible nature management. The measures provided for in the Directive are shown in Fig. 3.1. The processes of environmental impact assessment are presented in Fig. 3.2.

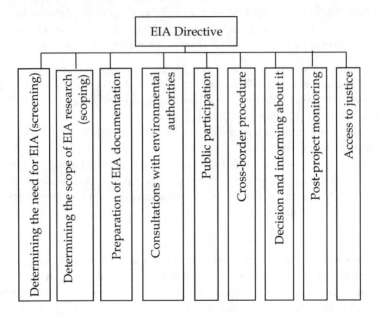

Fig. 3.1 Measures provided by the EIA Directive

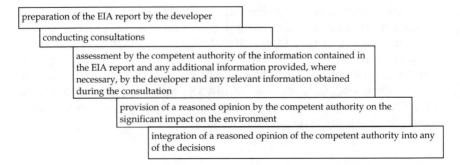

Fig. 3.2 Environmental impact assessment processes stipulated in the EIA Directive

3 Environmental Impact Assessment of the Planned Activity of Aviation Transport

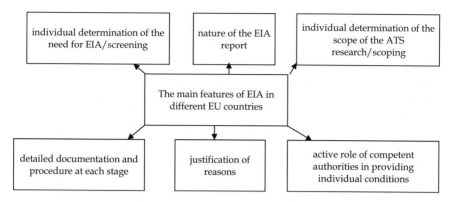

Fig. 3.3 The main features of EIA in different EU countries

The main features of EIA in different EU countries are presented in Fig. 3.3.

The implementation of international requirements in the legislation of Ukraine contributes to the introduction of a European approach to the formation of the domestic regulatory framework to identify the nature, intensity, and degree of danger of the impact of economic activity.

The European approach to environmental impact assessment, outlined in EU Directive 2011/92/EU and amendments made in 2014, is the basis of environmental legislation of Ukraine and the Law of Ukraine "On Environmental Impact Assessment" of May 23, 2017, No. 2059-VII, which is included in main priorities of the program document "Eastern Partnership – 20 expected achievements by 2020" (EBRD 2019).

Implementation of EIA in accordance with EU legislation and standards, compliance with the appropriate conventions (Espoo Convention, Aarhus Convention) and applicability in construction, and repair and operation of transport facilities are very important tasks and contribute to better governance for sustainable development.

Annex XXX (Environment) of the Association Agreement between Ukraine and the European Union requires the following: "Ukraine undertakes to gradually approximate its legislation to... Directive 2011/92/EU on the assessment of the effects of certain public and private projects on the environment" (EBRD 2019). Environmental impact assessment (EIA) in the construction, operation, and repair of transport facilities is a procedure that ensures that the environmental consequences of decisions are taken into account before making such decisions. It provides for the systematic collection and analysis of information on the project's environmental impact by the developer to enable the responsible authority to decide whether and how such a project should be implemented. The issue of EIA implementation is especially relevant for the development and implementation of large infrastructure projects for construction, operation, and repair of transport facilities, during which there is a significant impact on all components of the biosphere: atmosphere, hydrosphere, lithosphere, and biological diversity

The task of the Report on Environmental Impact Assessment of projects on construction, operation, and repair of transport facilities is to create a picture of a common understanding of the requirements, regulations, and issues related to the implementation of the planned construction/reconstruction/operation of the transport facility (Zub et al. 2019). The document is designed to determine and forecast the degree of environmental and socioeconomic impact of construction/reconstruction/operation of the transport facility.

The list of information that must be displayed, studied and detailed in the EIA report, is determined in Art. 6, clause 8 of Art. 4, and clause 5 of Art. 3 of the Law on EIA and in Annex 4 "EIA report" to the Procedure for submission of documents for provision of opinion on environmental impact assessment and financing of environmental impact assessment of construction/reconstruction/operation of transport facility (Pro 1991).

It is recommended to determine the scope of research and the level of detail of information in the EIA report to the business entity separately in each case, taking into account the requirements of special legislation (environmental protection, health, sanitary and epidemiological well-being, urban planning, industrial safety, occupational safety, etc.). In accordance with clause 2.5 of Art. 6 of the Law of Ukraine on EIA, description and assessment of possible environmental impact of the planned activity should include analysis of the magnitude and scale of such impact, nature (if available – transboundary), intensity and complexity, probability, expected onset, duration, frequency, and inevitability of impact.

Therefore, international and domestic legislation outlined the requirements for environmental impact assessment and the procedure for its implementation. To directly assess the environmental impact of projects on construction and reconstruction of air transport facilities, it is necessary to determine the appropriate criteria and the method of their quantitative assessment.

3.3 Evaluation Criteria and Indicators of Environmental Impacts of Transport Facilities (TFs)

In the process of assessing the impact of TFs on the environment, conduct an assessment of the impact on the following components of the natural and social environment: climate, air, geological environment, aquatic environment, soil, flora and fauna, man, material values, and cultural heritage. To select the criteria for assessing the impact of the project on the environment, the principles and procedures presented in ISO 21929-1, ISO 15392, DSTU ISO 14040, DSTU ISO 14020, DSTU ISO 14021, DSTU ISO 14024, and DSTU ISO 14025 are applied. In the case of construction/operation/reconstruction of a transport facility, the scope of research and the level of detail of information should take the following into account (Olekh et al. 2013; Nieviedrov et al. 2020; Khrutba and Nieviedrov 2020; DSTU 2020):

3 Environmental Impact Assessment of the Planned Activity of Aviation Transport 43

- Specific character of the transport structure
- Peculiar features of the planned location
- Planned types and scope of use of natural resources
- Peculiar features of planned constructions and technical solutions
- Planned economic characteristics

All these determine the potential levels of impact of the transport structure (both positive and negative) on the natural environment and the socioeconomic component of the planned activities.

DSTU 9060:2020 "Environmental Impact Assessment. Transport Structures. Assessment Criteria and Environmental Impact Indicators" sets out the criteria for environmental impact assessment and indicators of transport infrastructure facilities, namely, roads, airfield runways, hydraulic structures, and other (DSTU 2020).

Quantitative assessment of environmental impact is carried out for all types of planned activities (or groups of current factors), which include technological processes of construction/reconstruction/operation of TFs and have an impact on the environment.

For each type of planned construction/reconstruction/operation of TFs that have a primary (or direct) impact on the environment, the criteria for environmental impact assessment are determined in accordance with (Olekh et al. 2013; Nieviedrov et al. 2020).

The main aspects of the environmental impact of the TF are as follows:

- Impact on the quality of the surface layer of atmospheric air
- Volume of nonrenewable resources consumption
- Impact on the quality of the aquatic environment
- Indicator of waste management efficiency
- Impact on the quality of land resources
- Impact on the quality of the geological environment
- Physical factors influencing the environment
- Impact on flora and fauna, protected objects
- Impact of the TF on the social environment
- Impact of TFs on the man-made environment

Basic criteria, indicators, and main impacts of air transport on the environment are given in Table 3.1.

To assess the criteria and indicators of EIA of the planned activities for construction, operation, repair, reconstruction of TF the interaction between the types of planned activities (or groups of current factors) and environmental components (environmental characteristics) is established.

Quantitative evaluation of EIA criteria and indicators is performed using a combined approach by means of the improved Leopold matrix and its further study using the Harrington function. Other complementary methods can be used to conduct EIA: the method of combined map analysis, flow chart system method, simulation method, method of expert groups, etc. (Olekh 2015; Rudenko et al. 2013; Leopold et al. 1971).

Table 3.1 Basic criteria, indicators, and main environmental impacts of air transport

Criteria	Indicator
Quality of the surface layer of atmospheric air	Mass concentration of pollutant in the surface layer of atmospheric air in a certain area of influence of the vehicle
	The content of ozone-depleting substances
	PG concentration in the surface layer of atmospheric air in a certain area of influence of the vehicle
	Mass concentration of solid pollutants (dust)
Resource saving. Energy saving	The use of metallurgical slag during the construction of TF
	Use of recycled plastic and/or rubber materials during the construction of TF
	Consumption of natural resources
	Environmental friendliness of transport
	Use of alternative energy sources
Quality of the aquatic environment	Use of ecological products (materials and equipment) that have passed the appropriate certification and have the appropriate labeling
	Concentration of pollutants in water bodies
	Oxygen consumption level (biochemical and chemical)
	Impact on the ecological state of the surface water massif
Waste management	The amount (volume) of waste generation
	Application of safe waste management technologies
Quality of land resources	Mass concentration of pollutants in soils
	Fragmentation of territories
	Preservation of green areas
Quality of geological environment	Influence on geological processes
	Influence on hydrological processes
Physical factors influencing the environment	Acoustic impact on the environment
	Influence of vibration
	Light pollution of the environment
	Electromagnetic influence
	Thermal pollution of the environment
	Radiation pollution
Biodiversity	Flora
	Fauna
	Nature reserves
Social environment	Transport accessibility of TF to the main life support facilities
	Proximity of TF to public transport
	Comfort of the territory occupied by TF or around it
	Cultural and historical value of the territory, impact on the local cultural and historical heritage of the territory
	Living conditions of the population in the area of influence of TF
Man-made environment	Impact of natural hazards during construction/operation/reconstruction of TF
	Influence of man-made hazards during construction/operation/reconstruction of TF

Adapted from DSTU (2020)

The Leopold matrix is a double causal link used in EIA. It is formed in the form of a table, which in rows contains a list of processes that have an impact on the environment at different stages of the life cycle of the planned activity, in columns – criteria and indicators of environmental impact as environmental characteristics (Akhnazarova and Hordeev 2003).

Harrington desirability function is a function that establishes a correspondence between the obtained values of environmental impact indicators and the values of the dimensionless scale of Harrington desirability, formed by the results of probabilistic estimates of the desired consequences of these impacts.

The examples of the best technologies for developing the necessary measures to prevent the possible risks and reduce the negative impact on the environment, which will be reflected in the report on environmental impact assessment of transport facilities, are presented in DSTU 9061:2020 "Environmental Impact Assessment. Transport facilities. Guidelines for preparing the Report on environmental impact assessment," in which there are also recommendations and background information (Rudenko et al. 2013).

Thus, the proposed criteria allow to quantify the environmental impact of the planned activities in the construction and reconstruction of air transport facilities. At the same time, independent assessment of environmental impacts using the improved Leopold matrix and its further study using the Harrington function require a large number of calculations. To simplify the procedure, the appropriate software has been developed – electronic service for independent environmental impact assessment (ES_EIA/ECEIA – electronic calculator environmental impact assessment) (Table 3.1).

3.4 Electronic Service for Independent Environmental Impact Assessment

The electronic service for independent environmental impact assessment allows to determine the quantitative assessment of the environmental impact of the planned activities of the business entity according to clearly established indicators and the method of their assessment. It provides for the processing and storage of information on environmental impact assessment, which is presented in the EIA report.

It is an information portal with a modern design and functionality for different categories of users, which meets modern trends and requirements and is easy to use (Fig. 3.4).

Fig. 3.4 General view of the main page of the service ES_EIA

3.4.1 Target Audience

- Business entities that plan activities subject to environmental impact assessment
- Design organizations and enterprises that ensure the implementation of planned construction or reconstruction activities
- Representatives of the interested public who take part in the discussion of the EIA report
- Employees of the Ministry of Environmental Protection and Natural Resources and appropriate departments of regional state administrations, which evaluate the EIA reports and provide the business entity with the opinion on the results of the assessment

3.4.2 Incoming Data

- Information related to the user's identification
- Type of project

3.4.3 We Choose

- Types of planned activities and facilities that may have an impact on the environment and are subject to environmental impact assessment
- The main production processes that provide the selected type of planned activities (Fig. 3.5)
- Criteria for environmental impact assessment (Fig. 3.6)

3 Environmental Impact Assessment of the Planned Activity of Aviation Transport

Fig. 3.5 Deployment of the tree of criteria for selecting processes in the service ES_EIA

Fig. 3.6 View of the guide for selecting criteria in the service ES_EIA

3.4.4 We Calculate

- Quantitative values of each selected criterion.
- We define the generalized influence on a separate element of environment of those processes, which directly influence this element, using the method of the improved Leopold matrix.
- Comprehensive assessment of the environmental impact of planned activities is determined by the desirability function of Harrington.

3.4.5 Results of Calculation

(proposals for making management decisions of the project to provide the business entity with an opinion on the results of the evaluation)

- The project is accepted.
- The project is accepted after minor refinement.
- The project is accepted after significant refinement and implementation of the necessary environmental safety measures; reevaluation is performed.
- The project is not accepted.

3.4.6 Additional Features

- Preservation of the obtained results
- Submission of analytical reports on the results of work

3.4.7 Tasks Solved by the Electronic Service

- Introduces a single unified system for measuring common indicators for all stakeholders, which provides a common understanding of the requirements and the results obtained
- Ensures full and convenient access to the calculation of quantitative values of indicators, which are presented as qualitative parameters in the EIA report
- Provides business entities that plan activities to be assessed for environmental impact, the opportunity to determine quantitative indicators of environmental impact at the stage of preparation of the EIA report
- Provides the appropriate structures of the Ministry of Environmental Protection and Natural Resources, the special-purpose departments of oblast state administrations that evaluate the EIA reports with information on quantitative indicators of environmental impact in the submitted reports, to provide the business entity with an opinion on the results of the assessment
- Provides complete, structured, and accurate information on the levels of impact on individual elements of the environment and a generalized comprehensive assessment of the impact of the planned activities on the environment to the representatives of the interested public who take part in the discussion of the environmental impact assessment
- Reduces the time spent on assessing the impact of planned activities on each component of the environment
- Ensures the possibility to store information and analyze the results in a personal electronic account
- Provides openness and transparency of the obtained results, the ability to reduce the human factor in the assessment of environmental impact assessment to prevent possible corruption

The general structure of the electronic service for independent environmental impact assessment EC_EIA includes four main blocks (Figs. 3.7 and 3.8).

3 Environmental Impact Assessment of the Planned Activity of Aviation Transport 49

1	**The block of the beginning of work** and input of the general information	2	**The database block** includes the following information databases
	entering information related to the User's identification; presentation of the necessary data for selection from the formed databases; program access counter.		types of planned activities and facilities that may have an impact on the environment and are subject to environmental impact assessment; main production processes that provide the selected type of planned activities; criteria for environmental impact assessment.

Fig. 3.7 General structure of the electronic service for independent environmental impact assessment EC_EIA (block 1, 2)

3	The calculation block	4	The block of results presentation
	calculation of the quantitative value of the selected criterion; determination of the generalized impact on the elements of the environment of the processes that have a direct impact on this element; determination of a comprehensive environmental impact assessment of planned activities.		quantitative assessment of impact on each element of the environment; quantitative assessment of impact of planned activities on the environment; proposals for making project management decisions; preservation of the obtained results; submission of analytical reports on the results of work.

Fig. 3.8 General structure of the electronic service for independent environmental impact assessment EC_EIA (block 3, 4)

Navigation on the electronic service is simple and intuitive. The information is available to all categories of users. The interface language is Ukrainian.

Therefore, with the help of the electronic service, each of the interested parties can calculate the impact of the planned activities on the environment automatically. The electronic service simplifies the procedure for calculation of EIA and increases its objectivity and accessibility for understanding of the results obtained for all interested parties. The developed service allows to conduct a quantitative assessment of environmental impact and decides on the possibility of carrying out planned activities of the business entity. We will assess the environmental impact of a fragment of the project of reconstruction of the flight zone of SE "Boryspil International Airport."

3.5 Project "Reconstruction of the Flight Zone No. 2 of SE 'Boryspil' International Airport"

The service was inspected on the basis of the project "Reconstruction of the flight zone No. 2 of SE IA 'Boryspil'," the main initial data of which are as follows:

1. Name of the object (transport facility). Flight zone No. 2 of SE IA "Boryspil."
2. Customer data. State Enterprise "Boryspil International Airport" (SE IA "Boryspil").
3. Type of work (construction, repair, reconstruction). In pursuance of the Decision of the National Security and Defense Council of Ukraine of July 20, 2015, enacted by the Decree of the President of Ukraine of September 04, 2015, No. 535/2015, and the Concept of Development of Boryspil International Airport until 2045, approved by the Resolution of the Cabinet of Ministers of May 08, 2019, No. 293-p on the basis of the State Enterprise "Boryspil International Airport," an international hub airport is formed.

SE IA "Boryspil" carries out design and survey works of the stage "feasibility study" for the construction object "Reconstruction of the flight zone No.2 of SE IA 'Boryspil'." The complex of works on the reconstruction of the flight zone envisages the modernization of the Kyiv/Boryspil airfield and its engineering and technical infrastructure.

4. *The cost of work.* The estimated cost of the construction object is UAH 11.5 bln.
5. *The period of project implementation (performance of works).* 2019–2027 (including design works (three-stage design); passing the examination; conducting tender procedures to determine the designer, investor, contractor; carrying out of construction works; commissioning; etc.).

The existing aerodrome surfaces are in unsatisfactory condition, operated for more than 40 years, which significantly exceeds the time between major repairs, unsatisfactory condition of drainage elements along existing surfaces, the existing design load PCN is 39, PCN value for the new runway No.2 must be at least 120.

Reconstruction of the flight zone No. 2 of airfield of the SE "Boryspil International Airport" (Kyiv) envisages construction of a new runway, new taxiways, expansion of existing platforms, installation of the latest engineering and technological equipment (light signal, radio navigation, meteorological station, emergency rescue station, etc.), and arrangement of other infrastructure facilities that are necessary for safe and technological operation of the airport. In particular, it is planned to increase the length of the new runway with artificial turf (P13PS-2) to 3800 m and equip it with a light signal system with MBland-177 according to 1I1-B category of ICAO, with MBland-357 according to one category of ICAO.

6. *Complex of preparatory works.*

- Dismantling of the existing aerodrome surfaces of the flight zone No. 2: runway, taxiways, aerodrome sidewalks, and existing passages at the aerodrome
- Dismantling of engineering networks that fall into the reconstruction zone (networks of fuel pipelines, water supply, drainage, communication cables, light signaling systems, etc.)
- Removal of tree and shrub vegetation
- Removal of the vegetative layer of the soil

3 Environmental Impact Assessment of the Planned Activity of Aviation Transport

7. *New construction of buildings and structures is envisaged*

- Artificial runway (AR-2), main taxiway, connecting taxiways and high-speed lanes
- Insulated parking place for aircraft (AC), parking place for testing aircraft engines, cargo aircraft parking places
- Systems of radio navigation and light signaling equipment
- Emergency rescue station (ERS)
- Rainwater collection and disposal systems from aerodrome sites and treatment facilities
- Transformer substations
- Fuel filing station (FFS)
- Patrol road, entrances and platforms
- Airport fences with security alarms and video surveillance
- Operating base for deicing (preparation of anti-ice liquid) and parking places
- Storage facilities for civil protection of fire equipment and the regular guard

8. *Reconstruction of existing buildings and structures is envisaged, in particular.*

- Transformer substations
- Fuel pipeline networks
- Platform "S" with the device for treatment of aircraft with anti-icing fluid

Construction works are expected to be performed in the conditions of the operating airport, with the termination of operation of the flight zone No. 2.

The total area of aerodrome surfaces is 1,064,105 m^2.

3.5.1 EIA According to Certain Criteria

As an example, the EIA of the part of the project, which involves the construction of a fuel filing station (FFS).

3.5.2 Selection of EIA Criteria

The EIA criteria are selected in accordance with DSTU Environmental Impact Assessment. Transport facilities. Evaluation criteria and environmental impact indicators for the construction of a fuel filing station (FFS) of the project "Reconstruction of the flight zone No. 2 SE 'Boryspil International Airport'" – construction of a fuel filing station (FFS).

The analysis of the peculiarities of certain technological processes and equipment that will be used for the construction of a fuel filing station of SE IA "Boryspil" allowed to identify those that have a primary (or direct) impact on the environment. The criteria for assessing their impact on the environment are given in Table 3.2.

Table 3.2 Criteria and indicators of environmental impact of the main processes of the stage of highway construction

Element of the environment	Criterion	Indicator
Air	Influence on the quality of the surface layer of atmospheric air	Mass concentration of a pollutant in the surface layer of atmospheric air for a certain period of time in a certain area of influence of TF
		Mass concentration of solid pollutants (dust)
Land	Mass concentration of pollutants in soils	
Waste	The amount (volume) of waste generation	Construction debris
		Remains of asphalt concrete mixture
		Residues of gravel and sand production
Water	Concentration of pollutants in water bodies	Suspended substances
		Petroleum products
	Oxygen consumption level (biochemical and chemical)	Chemical oxygen consumption
		Biochemical oxygen consumption
Flora and fauna	Impact on flora	Species diversity of populations, dominant, valuable, and protected species
	Impact on fauna	Species diversity of animal populations, dominant, valuable, and protected species
	Impact on protected sites	
Physical factors influencing the environment	Acoustic impact on the environment	

3.6 Determination of Criteria and Indicators for EIA

Mass concentration of pollutant in the surface layer of atmospheric air for a certain period of time in a certain area of influence of the TF.

Emissions of pollutants in the surface layer of atmospheric air are formed in each defined process by emissions of substances from engines of construction machinery (jib cranes, excavators, etc.), emissions of pollutants during transportation of materials, equipment, and workers.

Actual concentrations of pollutants in the construction area of a fuel filing station (Table 3.3).

The obtained results on the corresponding processes are entered into the Leopold matrix by means of a program (Fig. 3.9).

The results of impact assessment on individual components of the environment are determined according to the form (Fig. 3.10 and Table 3.4).

Analysis of Table 3.5 shows that impact on most components that affect the quality of the atmosphere is satisfactory. However, impact on the state of atmospheric air in terms of acoustic impact is significant, and some additional attention needs to be paid to the control of pollutant emissions at the boundary of the protective strip and solid particles (dust).

Table 3.3 Quantitative evaluation of the criterion "Mass concentration of pollutants in the surface layer of atmospheric air (on the verge of sanitary rupture)"

Pollutant	Environmental impact from pollutant emissions in the surface layer of atmospheric air on the verge of sanitary rupture									
	Danger of exposure	Law, regulation	Public opinion	Scope	Financial expenses	New technology	Time or duration of exposure	Ability to manage	Score of criterion	Intensity of influence
	R	L	P	S	F	Tech	Time	M	I_{Score}	ω_i
NO$_2$	3	2	3	2	3	3	3	2	38	3
C	2	2	2	2	3	2	2	2	48	2
CO	1	1	3	1	2	2	3	2	26	2
Benz (a) pyrene	3	2	3	3	3	2	3	3	57	4
C_mH_n	1	1	3	1	2	2	3	3	39	3
Inorganic dust (PM) SiO$_2$ 70–20%	2	1	3	2	1	1	3	2	26	2
Inorganic dust (PM) SiO$_2$ > 70%	2	2	3	3	3	1	2	2	32	2
Xylene	2	2	2	2	3	2	3	2	32	2

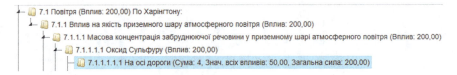

Fig. 3.9 Fragment of impact calculation according to Harrington in the service ES_EIA for the air criterion as an example

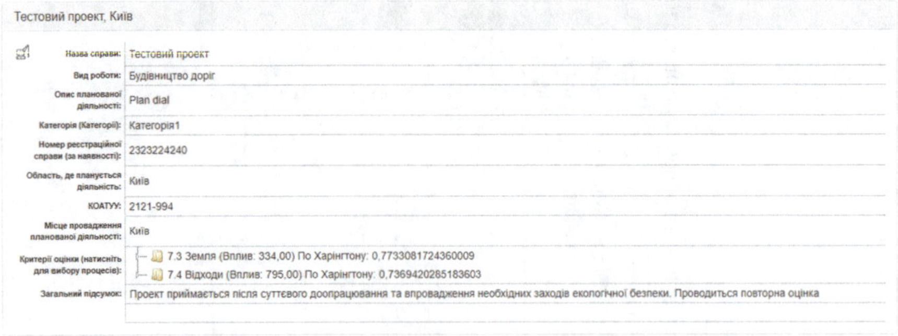

Fig. 3.10 Reporting of results according to the example in the service ES_EIA

Thus, the environmental impact assessment of the construction of fuel filing station (FFS) of the project "Reconstruction of the flight zone No. 2 of SE IA 'Boryspil'" – construction of fuel filing station (FFS) showed that the planned project activities will have almost no significant impact on the environment and can be performed after finalization of comments.

The list of measures taken at the site to reduce environmental impacts includes the following:

- Construction of closed local treatment facilities for deep surface water treatment from the territory of SE "Boryspil International Airport" (Kyiv), with the possibility of reuse of treated water for technical needs of the airport.
- Return surface waters will be treated on a sealed synergetic bioplateau with subsequent uniform drainage to the Ikva River.
- To ensure the requirements of aviation security; prevent unauthorized entry into the controlled area of the airport by unauthorized persons and wild and domestic animals; protection of the airport perimeter; the project provides for the equipment of the flight zone fence No. 2 with the perimeter security system.

Continuation of reconstruction and expansion of SE "Boryspil International Airport" and effective use of the country's transit potential with integration of its transport complex into the European and world transport and communication system will stimulate development of engineering and transport infrastructure of Ukraine and development of territories adjacent to transport corridors and the country in general.

Table 3.4 Quantitative assessment of the criterion "Acoustic influence of pollution sources in the surface layer of atmospheric air"

Noise source	Distance to the reference point	Environmental impact from acoustic impact in the surface layer of atmospheric air									
		Danger of exposure	Law, regulation	Public opinion	Scope	Financial expenses	New technology	Time or duration of exposure	Ability to manage	Score of criterion	Intensity of influence
		R	L	P	S	F	Tech	Time	M	I_{Scope}	ω_i
Crawler crane, jib RDK-250.2 (TAKPAF) – 25 t	3 m	3	3	3	2	3	2	3	2	38	3
	7.5 m	3	3	3	2	3	2	3	2	38	3
Jib crane GS 5363–25 t	3 m	3	3	3	2	3	2	3	2	38	3
	7.5 m	3	3	3	2	3	2	3	2	38	3
Crawler crane, electric MKGS-100 – 100 t	3 M	3	3	3	2	3	2	3	2	38	3
	7.5 m	3	3	3	2	3	2	3	2	38	3
Hitachi crawler crane KH-180-3, load-carrying capacity – 40 t	3 m	3	3	3	2	3	2	3	2	38	3
	7.5 m	3	3	3	2	3	2	3	2	38	3
KATO truck crane 500 E-3, load-carrying capacity – 50 t	3 m	3	3	3	2	3	2	3	2	38	3
	7.5 m	3	3	3	2	3	2	3	2	38	3
Pneumatic wheel crane "IVANOVETS" KS-35715, load-carrying capacity – 16 t	3 m	2	2	2	1	3	2	2	2	28	2
	7.5 m	2	2	2	1	3	2	2	2	28	2
Pneumatic wheel crane, load-carrying capacity – 10 t	3 m	2	2	2	1	3	2	2	2	28	2
	7.5 m	2	2	2	1	3	2	2	2	28	2
Bulldozer with capacity of 121 kW	3 m	2	2	2	1	3	2	2	2	28	2
	7.5 m	2	2	2	1	3	2	2	2	28	2
One-bucket crawler excavator – 0.65 m^3	3 m	3	3	3	2	3	2	3	2	38	3
	7.5 m	3	3	3	2	3	2	3	2	38	3

Table 3.5 Impact assessment on the state of atmosphere of individual components of pollution and negative impacts

	Components of impact on the state of atmosphere	Indicator value	Conclusion
1	Mass concentration of pollutant in the surface layer of atmospheric air (on the verge of sanitary rupture), mg/m^3	0.497894	Satisfactory (below average)
2	Mass concentration of pollutant in the surface layer of atmospheric air (at the boundary of the protective strip), mg/m^3	0.416654	Satisfactory (below average)
3	Mass indicator of solid pollutants (dust) emission, t	0.472793	Satisfactory (below average)
4	Mass indicator of pollutants emitted from bitumen and paint, t	0.328059	Satisfactory (below average)
5	Integral (total) indicator of acoustic impact (noise pollution)	0.647582	High
	The overall impact of the project on the state of atmosphere	*0.472597*	*Satisfactory (below average). The project is accepted for execution after finalization of comments*

3.7 Conclusions

International and domestic legislation outline the requirements for environmental impact assessment and the procedure for its implementation. To directly assess the environmental impact of projects for the construction and reconstruction of air transport facilities, it is necessary to determine the appropriate criteria and determine the method of their quantitative assessment. The main aspects of the environmental impact of TF are impact on the quality of the surface layer of atmospheric air; consumption of nonrenewable resources; impact on the quality of the aquatic environment; waste management efficiency indicator; impact on the quality of land resources; impact on the quality of the geological environment; physical factors influencing the environment; impact on flora and fauna, protected objects; impact of TF on the social environment; and impact of TF on the man-made environment. The proposed criteria allow to quantify the environmental impact of the planned activities during construction and reconstruction of air transport facilities. In addition, independent assessment of environmental impacts using the improved Leopold matrix and its further study using the Harrington function require a large number of calculations.

With the help of the electronic service for independent environmental impact assessment (EC_EIA), each of the interested parties can automatically calculate impact of the planned activity on the environment. The electronic service simplifies the procedure for calculating EIA and increases its objectivity and accessibility for understanding of the obtained results for all interested parties. The developed service allows to carry out a quantitative assessment of impact on the environment and decides on the possibility of carrying out the planned activities of the business entity. Environmental impact assessment of the construction of fuel filing station (FFS) of the project "Reconstruction of the flight zone No. 2 of SE IA 'Boryspil'" – the construction of fuel filing station (FFS) – showed that impact on most components that affect the quality of the atmosphere is satisfactory. However, impact on the state of atmospheric air from the point of view of acoustic impact is significant, and some additional attention needs to be paid to the issue of pollutant emissions at the boundary of the protective strip and solid particles (dust). The planned project activities will have almost no significant impact on the environment and can be performed after finalization of comments.

References

Akhnazarova S, Hordeev L (2003) Yspolzovanye funktsyy zhelatelnosty Kharrynhtona pry reshenyy optymyzatsyonnukh zadach khymycheskoi tekhnolohyy (Using Harrington's desirability function in solving optimization problems of chemical technology). Educational and methodical manual. M, RKHTU of DS Mendeleev

DSTU 9060:2020 (2020) Otsinka Vplyvu Na Dovkillia. Transportni Sporudy. Kryterii Otsinky Ta Pokaznyky Vplyvu Na Dovkillia (Environmental impact assessment. Transport facilities. Evaluation criteria and environmental impact indicators). State Standard of Ukraine, Kyiv

EBRD (2019) Methodology for the economic assessment of EBRD projects with high greenhouse gas emissions, Technical note. European Bank for Reconstruction and Development. https://www.ebrd.com/news/publications/institutional-documents/methodology-for-the-economic-assessment-of-ebrd-projects-with-high-greenhouse-gasemissions.html. Accessed 25 Jan 2019

Khrutba V, Nieviedrov D (2020) Metod kilkisnoho otsiniuvannia vplyvu na dovkillia v proektakh budivnytstva ta rekonstruktsii obiektiv krytychnoi infrastruktury (Method of quantitative assessment of environmental impact in projects of construction and reconstruction of critical infrastructure). Paper presented at the VII International Conference on "Project Management in Society Development". Kyiv, KNUBA, pp 252–255

Leopold LB, Clarke FE, Hanshaw BB, Others (1971) A procedure for evaluating environmental impact. U.S. Geological Survey, Washington, DC

Nieviedrov DS, Khrutba VO, Ziuziun VI, Barabash OV (2020) Formuvannia systemy kryteriiv otsinky vplyvu na dovkillia v proektakh budivnytstva ta rekonstruktsii obiektiv krytychnoi infrastruktury (Formation of a system of criteria for environmental impact assessment in projects of construction and reconstruction of critical infrastructure). Bull Natl Trans Univ Ser Tech Sci (K. NTU) 1(46):405–415

Olekh TM (2015) Rozrobka modelei tsilepokladannia ta metodiv pryiniattia rishen v proektakh na osnovi bahatovymirnykh otsinok (Development of goal-setting models and decision-making methods in projects based on multidimensional evaluations). Dis. Cand. of Technical Sciences. Odessa National Polytechnic University, Odesa, p 174

Olekh TM, Kolesnykova EV, Rudenko SV (2013) Ekolohycheskaia otsenka proektov (Environmental assessment of projects). Proc Odessa Polytech Univ 41:276–282. http://nbuv.gov.ua/UJRN/Popu_2013_2_52. Accessed 24 Feb 2013

Pro okhoronu navkolyshnoho pryrodnoho seredovyshcha (1991) (On environmental protection). "Law of Ukraine from 26.06.1991 №1268-XII".

Rudenko S, Olekh T, Hohunskyi D (2013) Model obobshchennoi otsenky vozdeistvyia na okruzhaiushchuiu sredu v proektakh (Model of generalized environmental impact assessment in projects). Manag Complex Syst Dev 15:53–60

Zub LM, Kostiushyn VA, Khrutba VO, Lievina HM, Sumskyi Ye D, Pylypovych OV, Kostiushyn Ye V, Matus SA, Yamelynets TS, Halaiko MB (2019) Pidhotovka zvitu z otsinky vplyvu na dovkillia pry budivnytstvi ta rekonstruktsii avtodorih: metodychnyi posibnyk (Preparation of a report on environmental impact assessment in the construction and reconstruction of roads: a guide) Kyiv, p 108

Chapter 4
Key Aspects of Sustainable Development Toward Spent Lithium-Ion Battery Recycling

Lina Kieush, Andrii Koveria, Andrii Hrubiak, and Serhii Fedorov

Nomenclature

EVs   Electric vehicles
RoHS   Restriction of Hazardous Substances
NCM   Nickel-cobalt-manganese
XRF   Fluorescence

4.1 Introduction

The rapid growth of the use of lithium-based batteries used in portable electronic devices, electric vehicles (EVs), and storage of excess energy requires a significant amount of metal resources, such as lithium, cobalt, manganese, nickel, and others. At the same time, LIB production is an energy- and resource-intensive technology, the

---

L. Kieush (✉)
Ukrainian State University of Science and Technologies (Former National Metallurgical Academy of Ukraine), Dnipro, Ukraine

A. Koveria
Dnipro University of Technology, Dnipro, Ukraine
e-mail: KoveriaA@nmu.org.ua

A. Hrubiak
Institute of Metal Physics, National Academy of Science, Kyiv, Ukraine

S. Fedorov
Ukrainian State University of Science and Technologies (Former National Metallurgical Academy of Ukraine), Dnipro, Ukraine

Thermal & Material Engineering Centre, Dnipro, Ukraine

Iron and Steel Institute of Z.I.Nekrasov of NAS of Ukraine, Dnipro, Ukraine

© The Author(s), under exclusive license to Springer Nature Switzerland AG 2023
S. Boichenko et al. (eds.), *Sustainable Transport and Environmental Safety in Aviation*, Sustainable Aviation, https://doi.org/10.1007/978-3-031-34350-6_4

improvement of which requires solutions based on the principles of sustainable development.

Batteries act as energy storage in EVs, and more than 34 million EVs (hybrid, plug-in hybrid, and battery electric vehicles) are expected to be sold in 2030, according to the base case scenario. They also can be an energy buffer in the power system, supporting the integration of renewable energy generation as a major base source (WEF 2019). Batteries are a key technology to decarbonize transport and support decarbonization in the power sector. The challenge is enormous: to get on track for the Paris Agreement 2 °C target, the transport and power sectors have a joint remaining carbon budget until 2050 of 430 GtCO$_2$e. Without batteries, this budget will be used up by 2035 and with batteries in the base case by 2040.

Electric cars are an innovative topic around the world. There is not a single major automotive company that has not invested significantly in development in this area. Factors of this approach are, firstly, that transportation services using electricity are much cheaper than those using hydrocarbon fuels. In Ukraine, due to the relatively cheap electricity so far, the ratio of travel expenses for 100 km of fuel to electricity is 5:1. Secondly, an important factor is the environmental friendliness of electric vehicles. Fossil resources (gas and coal) are still mostly used to generate electricity for charging LIBs. Nevertheless, the place is gradually being replaced by alternative energy.

Ukraine takes a special place in world trends. Despite the presence of a small share of electric cars (up to 2%), our country, for the fifth year, has a steady growth in sales of these cars at the level of market leaders – China and the United States. It is important to note that the vast majority of domestic electric cars (over 90%) are used cars whose batteries have already partially lost capacity. According to forecasts, the trend toward a large number of used cars in the Ukrainian market will continue in the future.

Meanwhile, an electric car's battery is a cell that rapidly inactivates. This is due to the fact that, with charging and discharging of LIB, there are irreversible destructive processes, in particular, the structural degradation of electrode materials, their oxidation, the emergence of a film on the surface of the carbon anode, etc. The number of charging-discharging cycles of a typical electric car battery is 500–1000 cycles on average. Depending on the discharge depth, the maximum lifetime of the batteries does not exceed 8 years. However, electric cars, which are imported to Ukraine, already have an average of 5 years of experience, which will lead to the accumulation of a large number of used batteries that require disposal.

Recycling the used LIBs to extract valuable materials for further use for the production of batteries is one of the key approaches in modern world practice. Issues of recycling spent LIBs are actively studied, but the best solution is in scientific research. There is significant potential for improving the recovery of valuable materials such as lithium, cobalt, nickel, copper, and carbon.

In (Chang-Heum and Seung-Taek 2019), authors used the technique of recycling, which is based on the difference in the solubility of materials in several solvents at different temperatures. Lithium and cobalt compounds were successfully recovered from spent electrodes. The obtained high-purity materials were compared with commercial materials to provide the required physical and chemical properties. It is observed that the electrochemical performance of the electrode material prepared from the purified material is similar to that of the commercially available electrode material. Besides, the recycling process is shown to be environmentally and economically advantageous.

Xiao et al. (2017) proposed an integrated treatment process for large lithium batteries, according to which the binder used in LIBs, graphite, and $LiMn_2O_4$ was first removed by mechanical separation. The results show that mixed electrode materials can be in situ converted into manganese oxide (MnO) and lithium carbonate (Li_2CO_3) at 1073 K for 45 min. In this process, the binder is evaporated and decomposed into gaseous products which can be collected to avoid disposal cost. Finally, 91.30% of Li resource as Li_2CO_3 is leached from roasted powders by water, and then high value-added Li_2CO_3 crystals are further gained by evaporating the filter liquid. The filter residues are burned in air to remove the graphite, and the final residues as manganous-manganic oxide (Mn_3O_4) are obtained.

Sattar et al. (2019) focused on sulfuric acid leaching of $LiNi_xCo_yMn_zO_2$ cathode material for resource recovery of valuable metals from spent LIBs. The process parameters, viz., pulp density, acid concentration, the dosage of reducing agent (i.e., H_2O_2), time, and temperature, have been optimized for leaching of cathode powder (of weight composition: 7.6% lithium, 20.48% cobalt, 19.47% manganese, and 19.35% nickel). The maximum 92% lithium and nickel, 68% cobalt, and 34.8% manganese could be leached while leaching a 5% pulp density in 3.0 M H_2SO_4 without H_2O_2 at 90 °C. Leaching efficiencies of metals were found to be increased within 30 min and reaching to >98% by adding 4 vol% H_2O_2 even at a lower temperature, 50 °C. Thereafter, selective precipitations of manganese and nickel were carried out from leach liquor using $KMnO_4$ and $C_4H_8N_2O_2$, respectively. Subsequently, a two-stage solvent extraction using 0.64 M Cyanex 272 (50% saponified) at equilibrium pH 5.0 and O:A of 1:1 was employed for the recovery of a highly pure solution of $CoSO_4$. Finally, lithium could be precipitated at Li^+: Na_2CO_3 of 1.2:1.0.

The main production capacity of LIBs is in Asia (China, Japan, South Korea) and the United States – about 90% of the market – in particular, China occupies more than 60% of the market, but the use of batteries is global, and therefore the technology of their processing at existing production sites around the world has prospects. It should be noted that market research shows good competitiveness and export potential of valuable materials extracted from spent LIBs. The global LIB market size is estimated to grow from USD 44.2 billion in 2020 to USD 94.4 billion by 2025; it is expected to grow at a CAGR of 16.4%. Thus, according to forecasts, the market of LIBs will have an overall annual growth rate of 14%, while in the transport sector – by 60% by 2025. In turn, the cost of extracting resources from the salt brine of the seas and oceans and hard rocks is approximately 1800 and 5000 USD per ton, respectively. The estimated cost of obtaining valuable materials in the processing of spent LIBs is approximately 1130 USD per ton.

Notwithstanding, the total resources of lithium and other valuable materials recovered in the recycling process are still small compared to the number recovered from natural sources, and the reuse of materials that make up batteries should be continued due to limited resources for lithium. Additionally, other valuable metals, as well as environmental issues that accompany the recycling of these materials, given the rapid development of the use of LIBs in the world.

4.2 The Main Types of Power Sources Based on Lithium

4.2.1 Lithium Primary Power Sources with Solid Cathodes and Aprotic Electrolyte

The reducing agent is lithium; the oxidizing agents are metal oxides, sulfides, or fluorocarbon. Electrolytes are solutions of lithium salts ($LiClO_4$, $LiBF_4$, or $LiBr$) in aprotic solvents: propylene carbonate, dioxolane, γ-butyrolactone, tetrahydrofuran, dimethoxyethane, etc. Depending on the type of oxidizer used, the current source has a discharge voltage of approximately 1.5 V (CuO, CuS, FeS, Bi_2O_3, and FeS_2) or 2,5–3,2 B (MnO_2, $(CF)n$, $Ag_2V_4O_{11}$, Ag_2CrO_4, $Cu_4O(PO_4)_2$, etc.). Lithium primary power sources have a higher capacity and specific energy, a wider operating temperature range, better performance at low temperatures, and a lower self-discharge rate in comparison with the same parameters of manganese-zinc power sources. Lithium current sources with a voltage of 1.5 V replace manganese-zinc batteries of the same standard size; current sources with a voltage of 2.5–3.2 V replace batteries of manganese-zinc cells. They are used in medical, consumer, industrial, and military electronics.

4.2.2 Lithium Power Supplies with Liquid or Dissolved Oxidant

These power supplies use sulfur dioxide (SO_2), which dissolves in an organic solvent, liquid thionyl chloride ($SOCl_2$), and sulfuryl chloride (SO_2Cl_2). The cathodes in the power supply are insoluble and are made of carbon materials deposited on an aluminum (for SO_2), nickel base, or stainless steel. The electrolyte in the cell of the lithium-sulfur dioxide system is $LiBr$ dissolved in acetonitrile, in cells with thionyl chloride and sulfuryl chloride – $LiAlCl_4$ in $SOCl_2$ or SO_2Cl_2 with additives. These power sources have a very high specific energy, high discharge rates, power density, a horizontal discharge curve, the ability to function at low temperatures (down to -55 °C), and long service life. The disadvantages include the relatively high cost, work under pressure, potential explosion hazard, and the presence of toxic components. They are used in those areas where high specific energy and power are required, long-term storage, the ability to work at low temperatures (in space and military equipment, memory storage systems, and other devices).

4.2.3 Lithium Iodine Power Sources with Solid Electrolyte

The oxidizing agent is iodine dissolved in solid polyvinyl pyridine, and the electrolyte is solid LiI salt, the thickness of which is continuously increasing as a result of a current-forming reaction. These power sources can be stored for a very long time and have high specific energy and a wide range of operating temperatures but very low discharge rate and power density. They are mainly used in pacemakers and are manufactured for this purpose in a special D-shape.

4.2.4 Primary Lithium-Thionyl Chloride Power Supplies

The nominal voltage of primary lithium thionyl chloride power supplies is 3.6 V. The maximum capacity is 36 Ah in a standard DD package. $Li/SOCl_2$ system elements have the best energy density per unit volume characteristics among lithium primary power sources.

4.2.5 Coin Type Lithium Manganese Dioxide Battery

These batteries have a nominal voltage of 3.0 V. They also have a long shelf life (5–10 years). The loss of rated capacity is less than 2% per year at room temperature storage. The operating temperature range of CR series batteries is 20–60 °C.

4.2.6 Lithium-Polymer (Li-Pol) Batteries

Lithium-polymer (Li-Pol) batteries are the newest in lithium technology. With approximately the same energy density as conventional LIBs, lithium-polymer devices can be manufactured in a variety of plastic geometries that are unconventional for conventional batteries, thin enough to fill any free space. At the same time, the space utilization efficiency is increased by approximately 20%.

Li-Pol battery is structurally similar to lithium-ion but has a gel electrolyte. As a result, it became possible to simplify the design of the cell, since leakage of the gelled electrolyte is practically impossible. With the same capacity, a lithium-polymer battery is lighter than a LIB. This is another very important advantage of

Li-Pol batteries. Currently, the international company EEMB has a very large list of manufactured Li-Ion and Li-Pol batteries. Among Li-Pol batteries, LP052030-PCB-LD (capacity 230 mAh) and LP383454-PCB-LD (capacity 750 mAh) are very popular. They do not have the most extreme characteristics among the produced Li-Pol batteries. However, the combination of their parameters turned out to be the most demanded in the market. For LP05230-PCB-LD batteries, more than 400 charge/discharge cycles are guaranteed, and for LP383454-PCB-LD, more than 500 cycles. The number of cycles is given under the condition of charging with a current of 1C (where C is the battery capacity) and discharging at room temperature.

4.3 Lithium-Ion Batteries

Commonly, LIBs are used in modern mobile devices (laptops, mobile phones, etc.). This is due to their advantages in terms of volumetric energy density compared to nickel-metal hydride (Ni-MH) and nickel-cadmium (Ni-Cd) batteries.

However, it should be noted that Ni-Cd batteries have one important advantage, which is the ability to provide higher discharge currents and performance at very low ambient temperatures. These properties are not of primary importance when powering laptops or cell phones, but many devices consume high currents, for example, power tools, electric shavers, etc. Until now, exclusively Ni-Cd batteries have powered these devices. However, currently, especially in connection with the limitation of the use of cadmium under the Restriction of Hazardous Substances Directive 2002/95/EC (RoHS 1), research on the creation of batteries without cadmium and with a high discharge current has intensified. Because of the search for the best material for the cathode, contemporary LIBs are being transformed into a whole family of chemical current sources, significantly differing from each other both in energy consumption and in the parameters of charge/discharge modes. This, in turn, requires a significant increase in the intelligence of charge controller circuits. Otherwise, it may damage (including irreversible) as a battery and powered devices.

LIBs of the international manufacturer of EEMB batteries of the LIR series is produced in two types of cases, namely, in tablet (another name is push button) and cylindrical. The main parameters and appearance of the batteries of this series are given in Tables 4.1 and 4.2.

The maximum capacity of prismatic LIBs is 1800 mAh at a nominal voltage of 3.7 V and a case weight of about 41.2 g. The capacity of button cell batteries reaches 200 mAh with a case diameter of 30 mm and a height of 4.8 mm. The popular LIR18650 has a maximum capacity of 2100 mAh among EEMB cylindrical LIBs with a weight of about 45.0 g. The nominal voltage of cylindrical LIBs is 3.7 V.

An example of a typical LIB is shown in Fig. 4.1. A LIB consists of electrodes (cathode material on aluminum foil and anode material on copper foil) separated by a porous separator impregnated with an electrolyte. The package of electrodes is placed in a sealed case, and the cathodes and anodes are connected to the current

Table 4.1 Parameters of lithium-ion batteries in button-type cases

Type	U_r, V	Rated capacity, mAh	Recommended charge current, mA DC	Pulse current	Dimensions, mm Diameter	Height	Weight, g
LIR2016	3.6	20 ± 5	1C, мA	2C, мA	20.0	1.6	1.9
LIR2025		25 ± 5				2.5	2.5
LIR2032		45 ± 5				3.2	3.1
LIR2450		120 ± 10			24.0	5.0	5.2
LIR3048		200 ± 10			30.0	4.8	7.3

Table 4.2 Parameters of common LIBs in cylindrical cases

Type	U_r, V	Rated capacity, mAh	Impedance, mΩ	Dimensions, mm Diameter	Height	Weight, g
LIR14500	3.7	800	≤80	14.1	48.5	20.0
LIR17500		1100		16.8	49.5	29.0
LIR18500		1300; 1400		18.2	48.5	33.0
LIR18650		1800; 2000; 2100; 2200			64.5	45.0

Fig. 4.1 Example of LIBs (LiNiCoMnO$_2$)

collector terminals. The package can be equipped with a safety valve that relieves the internal pressure in case of emergency or violation of operating conditions.

LIBs differ in the type of cathode material used. The charge carrier in a LIB is a positively charged lithium-ion, which can be incorporated (intercalated) into the crystal lattice of other materials (for instance, graphite, oxides, and metal salts) with the formation of a chemical bond, for example, into graphite with the formation of LiC$_6$, metal oxides (LiMnO$_2$), and salts (LiMn$_R$O$_N$). The use of cobalt oxides allows

Table 4.3 Most commonly used lithium-ion battery cathode chemistries (Targray 2020)

Chemistry	Nominal, V	Charge, V limit	Charge and discharge C-rates	Energy density, Wh/kg	Applications
Cobalt	3.60	4.20	1C limit	110–190	Cell phone, cameras, laptops
Manganese (spinel)	3.7–3.80	4.20	10C cont. 40C pulse	110–120	Power tools, medical equipment
NCM	3.70	4.10	~5C cont. 30C pulse	95–130	Power tools, medical equipment
Phosphate	3.2–3.30	3.60	35C cont.	95–140	Power tools, medical equipment

batteries to operate at significantly lower temperatures and increases the number of discharge/charge cycles per battery. Spreading lithium-iron-phosphate batteries is due to their relatively low cost. LIBs are used in combination with a monitoring and control system (MCS) or BMS (battery management system) and a special charge/discharge device.

Due to their low self-discharge and a large number of charge/discharge cycles, LIBs are the most preferred for use in alternative energy.

The incremental improvement of manganese and phosphate over older technologies is very evident. Lithium cobalt oxide offers the highest energy density, yet its thermal stability and ability to deliver high load currents fall a little short.

Cathode materials are comprised of cobalt, nickel, and manganese in the crystal structure forming a multi-metal oxide material to which lithium is added. This family of batteries includes a variety of products that cater to different user needs for high energy density and/or high load capacity. The table below breaks down the most commonly used LIB cathode chemistries on the market into four groups: cobalt, manganese, nickel-cobalt-manganese (NCM), and phosphate.

Table 4.3 shows the most commonly used LIB cathode chemistries. An important aspect of the production of lithium-ion batteries is that they contain raw materials that are considered critical or part of the candidates classified as critical, as determined by the European Commission, having a high economic value and moderate supply risk. Cobalt is considered 1 of the 27 most important raw materials, while lithium, nickel, and aluminum are the candidate materials. The demand for lithium carbonate is projected to increase from 265,000 tons in 2015 to 498,000 tons in 2025. Thus, it indicates that lithium will be more in short supply in the commodity market after 2023. The huge gap between market supplies and demand leads to a steady rise in prices for lithium carbonate and metals (cobalt, nickel). To avoid the risk of supply and reduce the cost of production, it is necessary to ensure the recovery of valuable metals from all possible resources.

The world example of pyrometallurgical processing is the Belgian plant (Umicore), which is the largest of its kind in Europe. Processing is based on a process that focuses on cobalt, nickel, and copper. A significant disadvantage of the process is the complete loss of individual components, in particular graphite material. Experts consider this process economically and ecologically unjustified.

Toxco technology (Canada) and Recupyl technology (France) are based on the hydrometallurgical process of recycling spent LIBs. Toxco technology involves pretreatment of batteries, separation of components, leaching, solution purification, and lithium deposition. Recupyl technology uses a combination of physical and chemical processing steps to produce lithium carbonate. The presented methods allow removing almost all elements, but a significant disadvantage of these methods is the low manufacturability of the process, which is associated with the need to use aggressive acids. Available commercial technologies focus mainly on extracted cobalt, lithium, aluminum, and copper with complete loss of graphite. Thus, the development of technology that allows recycling spent LIBs, regardless of their composition, to obtain precious metals and graphite is relevant worldwide.

The production and recycling of LIBs is one of the fastest-growing sectors in the world and has significant potential for more sustainable development.

Ukraine has high prospects in the world market for the production and processing of LIBs, as it is characterized by reserves of graphite, nickel, lithium, and copper. Currently, the production of LIBs and their utilization in Ukraine are practically undeveloped. Almost all batteries entering the Ukrainian market end up simply being stored or ending up in landfills, which is a major environmental issue. This is due to the lack of both the infrastructure for collecting spent items and technologies for their processing. Thus, the development of environmentally friendly, resource-saving innovative solutions for the recycling of spent LIBs is of great importance for the production of scarce precious metals and graphite, reduces environmental pollution, and has significant prospects in Ukraine and the world.

The importance of recycling LIBs is as follows:

(a) Recovery of materials from spent LIBs can reduce the need to extract raw materials and transport these materials.
(b) Environmental benefits will be achieved by reducing raw material production. The production of raw materials (metals such as lithium), which is part of the batteries, leads to the formation of approximately half of greenhouse gas emissions from the production of batteries. Recovering these valuable materials will reduce carbon emissions compared to extracting these materials from primary sources. Recycling LIBs can reduce emissions by 1 kg of CO_2 per kg of battery. Battery recycling (disassembly and hydrometallurgical recycling) emits about 2.5 kg of CO_2 per kg of battery, and recycling of primary material emits about 3.5 kg of CO_2 per kg of battery. Additionally, the use of secondary materials in new products usually requires less energy than the production/extraction of primary materials.
(c) Recycling used LIBs can lead to greater investment opportunities for production capacity. Nowadays, the battery recycling industry is underdeveloped due to, in particular, the low number of batteries that have reached the end of their useful life. However, in 2030, approximately 1.2 million batteries are expected to reach the end date.
(d) Additional economic benefits will arise by creating jobs in the recycled lithium-ion battery recycling sector.

4.3.1 Disassembly of Lithium-Ion Batteries

Research on disassembling was carried out on spent battery from a Chevrolet Volt electric car (Fig. 4.2). The characteristics of the battery from a Chevrolet Volt electric car (2012 battery/module from EM Chevrolet Volt Rated voltage = 111 V/5kWh) are presented in Table 4.4. The battery contains 30 cells or 2 blocks of 12 cells and 1 block of 6 cells. Each cell has a nominal voltage of 3.7 V and a maximum voltage of 4.15 V – measured value 34 Ah. It is well suited for multiple applications including eBike, electric scooters, EV conversions, and solar/photovoltaic systems.

Before disassembling the battery, it was discharged. Further, for safety, the voltage was checked. Then, in a special sealed box, in an inert argon atmosphere, we manually disassembled the package with a metal scalpel (Fig. 4.3). When opening the cell with a scalpel, care must be taken not to damage personal protective equipment such as gloves. The cuts in the cell bag are made along the edges of the

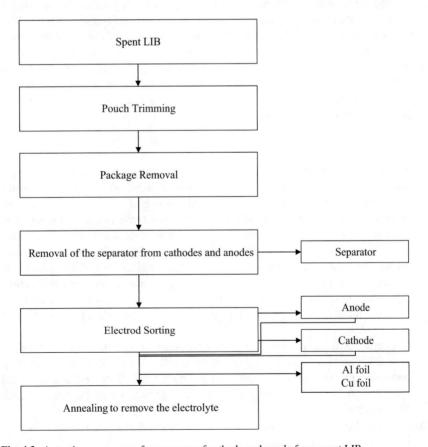

Fig. 4.2 A continuous process for recovery of cathode and anode from spent LIB

Table 4.4 Main characteristics of battery for Chevrolet Volt electric car

Characteristics	Parameters
Lithium battery module	$LiMn_2O_4/LiNiO_2$
Direct current, A	4C = 140 A, short time 250 A
Length, cm	58
Width, cm	18 (top)/24 (bottom)
Height, cm	28
Weight, kg	47

Fig. 4.3 Spent LIB (cell). (**a**) The open module, showing the cell inside. (**b**) An open-cell pouch, showing the layers of electrode materials within

cell so as not to cut the stack of electrodes and current collectors; then this stack is separated into individual components using tweezers. This minimizes contact between the disassembly's gloves and the electrolyte-soaked components. After this procedure, all instruments are thoroughly cleaned. The components of a cell are physically divided into three groups, namely, anode, cathode (Fig. 4.3), and other materials. This physical separation minimizes the possibility of cross contamination during further processing. However, it should be noted that when the cells were opened, annealing was carried out at a temperature of approximately 200 °C to remove the electrolyte. Then the "black mass" of the electrode can be easily separated from the copper and aluminum foil sheets (Figs. 4.4 and 4.5).

After annealing the cathode, it was cleaned from the surface of aluminum foil and homogenized to a powder form in a ceramic mortar (Fig. 4.6).

After disassembly, the chemical composition (Table 4.5) of the studied samples was detected by X-ray fluorescence (XRF) analysis using the EXPERT 3 L precision analyzer (Nykyruy et al. 2019) with the following features: the detection limit of the elements for 100 s (from 12 Mg to 92 U) is $\leq 0.05\%$; the detector (SDD) is with a nominal statistical loading of the spectroscopic tract 52,000 1/s; and the resolution of the detector (for $K\alpha Mn$) at a nominal loading does not exceed 149 eV.

The chemical composition of the cathode according to the results of XRF is characterized by a content of manganese and nickel in the ratio of 3.2:1. Because of the processes of the cathode, recovering, and technological conditions for obtaining powdered cathode, there are impurities in the form of such elements as S, P, Al, Fe, Si, and Co.

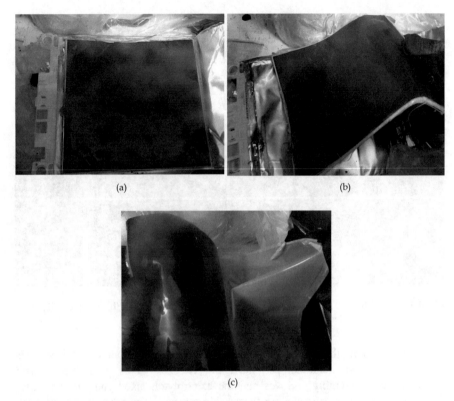

Fig. 4.4 Process of disassembly of spent LIB

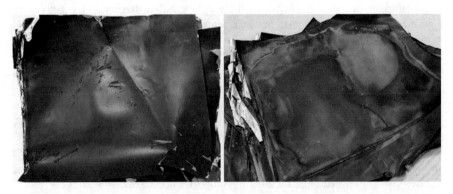

Fig. 4.5 Photographs of the cathode (left) and anode (right) after disassembly and annealing

Fig. 4.6 Powdered cathode from spent LIB

Table 4.5 Chemical composition of the cathode

Element	[%]
O	35.33
Al	0.62
Si	0.10
P	0.32
S	0.05
Mn	47.06
Fe	0.20
Co	1.77
Ni	14.55

4.4 Thermal Purification

To perform the thermal purification of valuable metals and graphite, a laboratory high-temperature furnace was used (Fig. 4.7).

The reactor is intended for the thermal processing of valuable metals and carbon materials at temperatures of up to 3000 °C under an inert gaseous environment. Its description is as follows:

- The furnace with the material cell equipped with electric heaters, thermal insulation, and main body
- Power supply source
- Inert gas supply source

The principal technical features of the reactor are as given: power, 5 kW; voltage, 220/12–30 V; current, 50–300A DC; and the maximum weight of a sample, 30 g. The type of heating system is resistive. The unit is equipped with a temperature control system.

Fig. 4.7 High-temperature unit for thermal treatment of valuable metals and carbon materials by TMEC

A typical cycle of thermal treatment consists of loading (5 min), drying (15 min), holding time (10 min), treatment (optional duration), and cooling down (30 min).

Thermal purification of valuable metals and carbon materials (particle size less than 0.2 mm) was conducted via the lab reactor at 3000 °C (Kieush et al. 2019, 2022; Fedorov et al. 2020). During the experimental study, the treatment of each sample lasted for approximately 5 min under steady conditions.

The thermal treatment technology provides fast purification of valuable metals and carbon materials. It takes several minutes to reduce the total ash content of the material to less than 0.01%. That presents a significant advantage over Acheson's conventional chemical processes. Practically, such performance indicators as the operating temperature and the treatment duration can be reached only via electro-thermal purification technology applied as based on a fluidized bed, so-called EFB technology (Fedorov et al. 2015).

Within this study, an EFB lab reactor, capable of purifying the valuable metals and carbon material intended for carbon production at continuous mode, is being developed. Regarding the environmental issues, it is also important to develop a new engineering concept of an industrial purification unit based on electric supply from renewable energy sources that will significantly decrease operating costs.

4.5 Conclusions

LIBs are increasingly used as energy storage devices for vehicles, electronics, renewable sources, etc. The scope of their application is constantly increasing, due to their special properties, namely, high energy density, long duty cycle, low self-discharge, and safe handling. However, the large-scale use of LIBs raises more issues related to the extraction and preparation of raw materials, the production of batteries, and their disposal. These issues directly relate to the environment, mining and processing, and environmental management and therefore require a serious scientific approach, taking into account the environmental, technical, economic, and social characteristics of production, use, and disposal of LIBs, which are in line with the key principles of sustainable development. After the end of battery life, valuable metals require special treatment because their accumulation leads to environmental consequences and is not justified from an economic and resource-saving point of view, given the presence of a large number of valuable materials that can be reused for LIBs or other productions.

In this chapter, the approach involving the extraction of cathode and anode materials for the subsequent recovery of valuable materials has been studied and performed. Thermal purification as a method for the recovery of valuable metals and carbon material via the lab reactor at 3000 ° C has been proposed.

References

Chang-Heum J, Seung-Taek M (2019) Efficient recycling of valuable resources from discarded lithium-ion batteries. J Power Sources 426:259–265

Fedorov S, Rohatgi U, Barsukov I et al (2015) Ultrahigh-temperature continuous reactors based on Electrothermal fluidized bed concept. J Fluids Eng 138(4):044502. https://doi.org/10.1115/1.4031689

Fedorov S, Kieush L, Koveria A et al (2020) Thermal treatment of charcoal for synthesis of high-purity carbon materials. Petrol Coal 62(3):823–829

Kieush L, Fedorov S, Koveria A et al (2019) The biomass utilization to obtain high-purity carbon materials. In: Problems of chemmotology. Theory and practice of rational use of traditional and alternative fuels and lubricants, pp 20–32. https://doi.org/10.18372/38222

Kieush L, Koveria A, Boyko M et al (2022) Influence of biocoke on iron ore sintering performance and strength properties of sinter. MMD 16:55–63. https://doi.org/10.33271/mining16.02.055

Nykyruy L, Ruvinskiy M, Ivakin E et al (2019) Low-dimensional systems on the base of PbSnAgTe (LATT) compounds for thermoelectric application. Physica E Low-Dimens Syst Nanostruct 106:10–18

Sattar R, Ilyas S, Nawaz H et al (2019) Resource recovery of critically-rare metals by hydrometallurgical recycling of spent lithium ion batteries. Sep Purif Technol 209:725–733

Targray (2020) Cathode materials for Li-ion battery manufacturers. https://www.targray.com/li-ion-battery/cathode-materials. Accessed 3 Nov 2020

WEF (2019) A vision for a sustainable battery value chain in 2030 unlocking the full potential to power sustainable development and climate change mitigation. Insight report, World Economic Forum, September 2019. http://www3.weforum.org/docs/WEF_A_Vision_for_a_Sustainable_Battery_Value_Chain_in_2030_Report.pdf. Accessed 3 Nov 2020

Xiao J, Li J, Xu Z (2017) Recycling metals from lithium-ion battery by mechanical separation and vacuum metallurgy. J Hazard Mater 338:124–131

Chapter 5
Green Technologies of Information Protection in Computer Networks of Electric Transport System

Halyna Holub, Ivan Kulbovskyi, Vitalii Kharuta, Olga Zaiats, Mykola Tkachuk, and Valentyna Kharuta

5.1 Introduction

Computer systems for monitoring and diagnosing the condition of energy facilities and subway systems allow to detect in advance the threat of damage to the main equipment and prevent the occurrence of abnormal situations, which in the future may cause accidents at these objects under investigation.

In order to continuously monitor the parameters of the operating modes of both electrical networks and other objects of the subway power supply system, the principle of synchronicity of the measured information is used, as well as the principle of unity (Stasiuk et al. 2012; Holub and Soloviova 2019), which will ensure the processing of information data to ensure the process of operational management of the control center. The basis of this process is intellectualization and informatization due to the integrated information space, which contains the parameters of the primary data obtained from certain system points and studied by data processing methods.

Information on the parameters of the primary data of the power supply modes of the system and the state of the subway electrical equipment undergoes the stage of intellectual processing to achieve the goal of obtaining results of the process of control and diagnostics of power facilities at the power supply distance. This

---

H. Holub · I. Kulbovskyi · M. Tkachuk
State University of Infrastructure and Technologies, Kyiv, Ukraine

V. Kharuta (✉) · O. Zaiats
National Transport University, Kyiv, Ukraine

V. Kharuta
College of Information Technology and Land Management of the National Aviation University, Kyiv, Ukraine

© The Author(s), under exclusive license to Springer Nature Switzerland AG 2023
S. Boichenko et al. (eds.), *Sustainable Transport and Environmental Safety in Aviation*, Sustainable Aviation, https://doi.org/10.1007/978-3-031-34350-6_5

information goes through the stage of archiving and storage in the original form of obtaining a certain period (Kulbovskyi et al. 2019; Fomin et al. 2019; Holub and Soloviova 2019).

During the monitoring process, a lot of data is received, which is stored, processed, and accumulated in the database. In this case, a necessary condition is the security of the server for the reliability of its operation and information security, which is possible due to the software and hardware of unauthorized access. In addition, for the timely transmission of information to higher levels of management, the server of local area network (LAN) provides information transmission through a router that connects the LAN to the regional or central network of the corporate system.

Analyzing the structure of the system of monitoring, control, and diagnostics of the parameters of the modes of electric power systems of the subway, we can conclude that to ensure reliability, the main object of the system is the main central server of the network. The work of the server provides information from the phasor measurement unit (PMU) on the parameters of electrical networks and objects of traction, traction-lowering, and transformer substations of special units of the metro power supply service, which forms the information data space (Lenkov et al. 2008; Holub and Soloviova 2019) and provides information exchange through the network of the central corporate system, commercial information processing, and the primary data of the emergency mode parameters. Therefore, the complexity of the process of receiving, processing, and transmitting information to the main central server of the network requires the protection of information. And since these are objects of increased complexity, it is necessary to ensure the protection of information of the system and each component of the system network, which can be achieved by models and methods of assessing the level of information protection.

5.2 Analysis of Literature Data and Problem Statement

According to the analysis of recent studies (Lenkov et al. 2008; Ihnatov and Huziy 2005; Borankulova and Tungatarova 2019), it is important to develop methods and models of information attack processes that allow you to assess the level of protection of computer systems from unauthorized access methods.

In (Stasiuk et al. 2012; Holub et al. 2020), the authors investigated the methods of algorithmic solutions for the formation of space from the primary information of the parameters of the modes of operation of electrical networks and presented mathematical models for processing this information. According to the fact that information is accumulated and information space is formed in the form of sets of multidimensional data arrays corresponding to the time, at all levels of the distributed computer network system.

The completeness of the description of the processes of attack on the simulated information can be ensured by using a fundamentally new concept of modeling. An urgent practical task is to develop new analytical models of information attack processes that can reflect the dynamics of information conflict and its protection (Ihnatov and Huziy 2005; Hryshchuk 2010).

In (Vorob'yov et al. 2007), the authors developed a model of information attack and information protection processes based on the mathematical apparatus of matrix-game theory, but the scope of this matrix-game model is quite limited, because the game in such a setting does not always have solutions in pure strategies. In decisions of mixed strategies, the level of security of the automated system is assessed a posteriori.

In the domestic scientific and technical literature, it is proposed to apply game theory for information security management (Ihnatov and Huziy 2005), which allowed to develop a quantitative approach, and this became the basis for further research, for example (Borankulova and Tungatarova 2019).

The application of methods of differential game theory to solve information security problems allows to describe the dynamics of processes occurring in information security systems during the attack on information and its protection and their random nature, as well as to adequately reflect the conflict of interests of information conflict (Ihnatov and Huziy 2005) – methods of unauthorized access and methods of information protection.

One of the promising mathematical tools are the methods of the theory of differential transformations of Academician of the NAS of Ukraine GE Pukhov (1986), which provide the ability to find solutions to information security problems in both real and accelerated time without losing the accuracy of the original mathematical model.

In (Hryshchuk 2010), the results of adaptation of methods of theories of differential games and transformations for the decision of applied problems of protection of the information are presented, namely, examples of development of new patterns of behavior in operating systems of detection of attacks and the differential game analysis of reliability of information protection system are considered. At the same time, the direction of further research is outlined – the expansion of the range of information protection problems solved by these methods.

Based on the above, to solve the problem of finding models of attack processes on multitasking server information, determining optimal strategies for allocating information protection resources, and determining the guaranteed level of information security, it is advisable to use the theory of differential games and transformations (Vorob'yov et al. 2007; Hryshchuk 2009; Voronko 2013).

5.3 The Purpose and Objectives of the Study

The aim of the work is to study mathematical models and methods for protection of computer networks of the subway power supply system from information attacks based on methods of differential game simulation of cyberattack and information protection (Hryshchuk 2009), which will provide optimal strategies for information protection in cyberattacks. To achieve this goal, the following tasks are defined:

- To determine in what states the information system can be.
- To develop mathematical models of information attack processes that can reflect the dynamics of information conflict and its protection.
- To carry out computer modeling of processes of attack on functionality of the server with the use of mathematical models of the information conflict.
- To determine the main optimal strategies for information protection in cyberattacks.

5.4 Development of a Mathematical Model of the Process of Attacking Information Attacks on the Distance Server of the Subway Power Supply System

The server is in working condition during the process of monitoring the parameters of the modes of electrical networks and equipment of the subway system in accordance with the time. But during information threats and attacks, it passes into other states (Hryshchuk 2009, 2010; Voronko 2013) and in accordance with the reliability of its work, with the corresponding probabilities, for some time T, which is equal to the duration of the interval of information attacks (5.1):

$$t \in [0, T]. \tag{5.1}$$

To compile the Kolmogorov-Chapman differential equations (Voronko 2013; Holub et al. 2020), we construct a graph model of the process of attacking information (Nguyen et al. 2018), and the probabilities of the states of the system are determined as follows:

$P_{F0}(t)$ – the probability of server failure: violation or failure to perform any of the provided functions

$P_{F1}(t)$ – the probability of server failure: violation or failure to perform any of the provided functions

$P_{F2}(t)$ – probability of communication failure with the upper level system

$P_{F3}(t)$ – the probability of failure of the administration functions of the database of emergency parameters

$P_{F4}(t)$ – probability of failure of commercial database administration functions

$P_{S0}(t)$ – the probability of providing the server with full functionality

$P_{S1}(t)$ – the probability of providing communication functions with the lower level system

$P_{S2}(t)$ – the probability of providing communication functions with the upper level system

5 Green Technologies of Information Protection in Computer Networks...

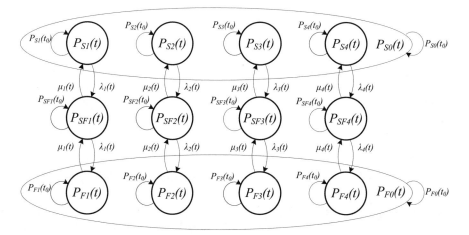

Fig. 5.1 Graph model of the process of information attack on the server of the distance of the power supply of the subway

$P_{S3}(t)$ – the probability of providing the functions of administration of the database of emergency parameters

$P_{S4}(t)$ – the probability of providing the functions of the administration of a commercial database

$P_{SF1}(t)$ – the probability of an attack on communication functions with the lower level system, under the influence of methods of information protection

$P_{SF2}(t)$ – the probability of an attack on communication functions with the top-level system, under the influence of information security methods

$P_{SF3}(t)$ – the probability of attack on the functions of the administration of the database of emergency parameters, under the influence of methods of information protection

$P_{SF4}(t)$ – the probability of attack on the functions of the administration of a commercial database, under the influence of methods of information protection.

Figure 5.1 shows a graph model of the process of information attack on the server of the distance of the power supply of the subway.

The nodes of the graph indicate the probabilities of states, and above the transition arrows that move the server from state to state, the intensity of the flows of protective actions $\mu_j(t)$ and information attacks $\lambda_i(t)$, which indicates their nonstationary nature, which is generally determined by functional time-parameter dependences (Vorobiev et al. 2017).

As can be seen from the above definitions of the states of the information system, the trouble-free operation of the server *S0* provides full functionality, which ensures the stay of the system in the following states *S1*... *S4* simultaneously, which are independent events. An *F0* server failure related to any of its features is an event. It is possible only when there is at least one of the independent failures *F1*... *F4*. Thus, for the probabilities of states *S0* and *F0*, we can write the following:

$$\begin{cases} P_{S0}(t) = \prod_{i=1}^{4} P_{Si}(t); \\ P_{F0}(t) = 1 - \prod_{i=1}^{4} (1 - P_{Fi}(t)). \end{cases} \qquad (5.2)$$

Based on the graph model (5.2), we write a system of differential equations that describes the dynamics of independent attacks on the functionality of a multitasking server:

$$\begin{cases} \dfrac{dP_{Si}(t)}{dt} = -\mu_i(t)P_{Si}(t) + \lambda_i(t)P_{SFi}(t); \\ \dfrac{dP_{SFi}(t)}{dt} = -(\lambda_i(t) + \mu_i(t))P_{SFi}(t) + \lambda_i(t)P_{Si}(t) + \mu_i(t)P_{Fi}(t); \\ \dfrac{dP_{Fi}(t)}{dt} = -\mu_i(t)P_{Fi}(t) + \lambda_i(t)P_{SFi}(t), \quad i = 1..4. \end{cases} \qquad (5.3)$$

The probabilities of the states of the information system should take into account the rationing condition, which is set as an expression for a complete group of events:

$$P_{Si}(t) + P_{Fi}(t) + P_{SFi}(t) = 1, \quad i = 1\ldots4. \qquad (5.4)$$

Thus, the integrated consideration of cyberattack processes on the server should be supplemented (5.3) by Eqs. 5.2 and 5.4, resulting in a mathematical model (5.5) of the information conflict in question:

$$\begin{cases} P_{S0}(t) = \prod_{i=1}^{4} P_{Si}(t); \\ P_{F0}(t) = 1 - \prod_{i=1}^{4} (1 - P_{Fi}(t)); \\ P_{Si}(t) = 1 - P_{Fi}(t) - P_{SFi}(t); \\ \dfrac{dP_{SFi}(t)}{dt} = -(\lambda_i(t) + \mu_i(t))P_{SFi}(t) + \lambda_i(t)P_{Si}(t) + \mu_i(t)P_{Fi}(t); \\ \dfrac{dP_{Fi}(t)}{dt} = -\mu_i(t)P_{Fi}(t) + \lambda_i(t)P_{SFi}(t), \quad i = 1..4. \end{cases} \qquad (5.5)$$

The following initial conditions are valid for the corresponding probabilities of states:

$$\begin{cases} P_{S0}(t_0) = 1, P_{F0}(t_0) = 0; \\ P_{SF1}(t_0) = P_{SF2}(t_0) = P_{SF3}(t_0) = P_{SF4}(t_0) = 0. \end{cases} \qquad (5.6)$$

The system of Eq. 5.5 allows to determine the distribution of probabilities of the server in each state during the information conflict (Ihnatov and Huziy 2005; Voronko 2013) taking into account the intensity of the flow of attacks and protective actions of information security methods.

5 Green Technologies of Information Protection in Computer Networks... 81

Due to the fact that in real conditions the change of strategies of the parties is caused by many factors, which in most cases cannot be taken into account, let's say that the strategies of information protection methods (Song et al. 2015; Khoroshko and Morzhov 2004; Korobiichuk et al. 2018) are distributed according to linear laws of general form:

$$\lambda_i(t) = \lambda_i \cdot t,$$

$$\mu_j(t) = \mu_j \cdot t, \tag{5.7}$$

where t is the time argument according to (5.1); i, j is the number of transitions between states as a result of successful attacks of the violator and as a result of the methods of information protection; and λ_i and μ_j are parameters for approximating the laws of distribution of method strategies.

The resources of defensive actions μ_j and information attacks λ_i are subject to restrictions of the following form:

$$\lambda_{i\,\min}(t) \le \lambda_i(t) \le \lambda_{i\,\max}(t)$$

$$\mu_{j\,\max}(t) \le \mu_j(t) \le \mu_{j\,\max}(t), \tag{5.8}$$

where, respectively, $\lambda_{i\,\min}(t)$, $\mu_{j\,\min}(t)$ is the minimum and $\lambda_{i\,\max}(t)$, $\mu_{j\,\max}(t)$ is the maximum intensities of information attacks and defensive actions.

The intensities of players' actions $\lambda_i(t)$, and $\mu_j(t)$, which determine the resources of the parties to the game, lie within closed sets $\Lambda \in V_\lambda$ and $M \in V_\mu$, which in turn are limited in Euclidean spaces E_λ and E_μ, accordingly (Hryshchuk 2010; Korobiichuk et al. 2019; Demin 2017).

During an information conflict, players try to achieve opposite goals. The player, or method of protecting information protected from attacks, tries to ensure the functional stability of the server by guaranteeing its security and the attacking player – to achieve server denial of service for certain functionality (Hryshchuk 2009, 2010; Voronko 2013). To do this, the player who defends himself from the attack tries to ensure the least loss by choosing such a strategy $\mu_i(t)$, that minimizes the fee, provided it is maximized by another player, which is formalized in the following form:

$$\min_{\mu_j(t) \in V_\mu} \max_{\lambda_i(t) \in V_\lambda} I\big(t, P_{F0}(t), \lambda_i(t), \mu_j(t)\big). \tag{5.9}$$

The violator maximizes the fee while minimizing their own losses during the attacks:

$$\max_{\lambda_i(t) \in E_\lambda} \min_{\mu_j(t) \in E_\mu} I\big(t, P_{F0}(t), \lambda_i(t), \mu_j(t)\big). \tag{5.10}$$

If the boards of both parties (5.9) and (5.10) are equal, the relation is as follows:

$$\min_{\mu_j(t)\in E_\mu} \max_{\lambda_i(t)\in E_\lambda} I = \max_{\lambda_i(t)\in E_\lambda} \min_{\mu_j(t)\in E_\mu} I = I^*\left(t, P_{F0}^{opt}(t), \lambda_i^{opt}(t), \mu_j^{opt}(t)\right). \qquad (5.11)$$

The strategies of players $\lambda_i^{opt}(t)$ and $\mu_j^{opt}(t)$ are optimal for the game in question, and $P_{F0}^{opt}(t)$ are the optimal trajectory, which is calculated from the system of Eq. 5.5 by criterion (5.11). Deviation of any of the players from their optimal strategy leads to corresponding losses in the board (5.11).

Therefore, the guaranteed level of server security is achieved by players choosing the optimal strategies $\lambda_i^{opt}(t)$ and $\mu_j^{opt}(t)$:

$$I^*\left(t, P_{F0}^{opt}(t), \lambda_i^{opt}(t), \mu_j^{opt}(t)\right) = I^G, \qquad (5.12)$$

where I^G is the price of the game, which determines the guaranteed level of security of the server.

In the conditions of dynamic information conflict, the game board can be represented in an integral form (Hryshchuk 2010; Holub et al. 2020):

$$I = \frac{1}{T}\int_{t_o}^{T} P_{F0}(t)dt, \qquad (5.13)$$

with restrictions

$$0 \leq I \leq 1. \qquad (5.14)$$

Integration is carried out along the trajectory of the game from the initial moment of time $t = t_0$ to the end of the information conflict $t = T$.

5.5 Simulation of Information Processes of Attack Attacks on the Server of the System of Distance of Power Supply of the Subway

Modeling the process of attacking server functionality using (5.5) in analytical form is a complex mathematical procedure that requires the processing of large amounts of information in real and accelerated time.

To model the process of attacking information in real and accelerated time without losing the accuracy of the original model (5.5), the use of P-transformations is proposed (Pukhov 1986; Voronko 2013; Holub et al. 2020).

The use of P-transformations has the advantage of reducing the amount of calculations by numerical methods, which is achieved through the analytical

5 Green Technologies of Information Protection in Computer Networks... 83

capabilities of this operating method. Representation of the original model (5.5) in the field of images by the method of differential transformations allows to preserve the accuracy of the original model while excluding the time argument. As a result, modeling the process of attacking information is reduced to performing arithmetic operations in the image area.

P-transformations are called functional transformations of the following form (Holub et al. 2020):

$$X(k) = \frac{H^k}{k!} \left[\frac{d^k x(t)}{dt^k} \right]_{t=0} \Leftrightarrow x(t) = \sum_{k=0}^{\infty} \left(\frac{t}{H} \right)^k X(k), \qquad (5.15)$$

where $x(t)$ is the original, which is continuous, differentiable an infinite number of times, and limited, along with all its derivatives, the function of the real argument t; $X(k)$ is designation of the differential image of the original, which is a discrete (lattice) function of an integer argument, where $k = 0, 1, 2, \ldots$; and H is a scale factor that has the dimension of the argument t and is often chosen equal to the period of time on which the function is considered $x(t)$.

Differential images $X(k)$ are called differential T-spectra, and the values of the T-function $X(k)$ at specific values of the argument k are called discrete.

We translate the original model (5.5) by the method of P-transformations of the form (5.15) into the region of T-images, while the scale factor H is taken equal to the duration of information attacks T. Then, the system in the field of images will take the following form:

$$\begin{cases} P_{S0}(k) = \prod_{i=1}^{4} P_{Si}(k); \\ P_{F0}(k) = 1 - \prod_{i=1}^{4} (1 - P_{Fi}(k)); \\ P_{Si}(k) = 1 - P_{Fi}(k) - P_{SFi}(k); \\ P_{SFi}(k+1) = \frac{T}{k+1} [(-(\Lambda_i(k) + M_i(k))P_{SFi}(k) + \Lambda_i(k)P_{Si}(k) + M_i(k)P_{Fi}(k)], \\ P_{Fi}(k+1) = \frac{T}{k+1} [(-M_i(k)P_{Fi}(k) + \Lambda_i(k)P_{SFi}(k)], i = 1..4. \end{cases}$$

$$(5.16)$$

where $P_z(k)$, $\Lambda_i(k)$, $M_j(k)$ is the differential images of the original functions $P_z(t)$, $\lambda_i(t)$, $\mu_j(t)$ accordingly.

Given the accepted assumptions that the strategies of players in the course of information confrontation change according to linear laws (5.7), when moving to the field of P-images, it is necessary to take into account the properties of T-products of differential images $\Lambda_i(k) \, P_z(k)$ and $M_j(k)P_z(k)$ (Pukhov 1986), one of the terms in which is a constant λ and μ multiplication by an integer degree of the independent variable T^m (with $m = 1$), in general:

$$\Lambda_i(k)P_z(k) = \lambda_i TP_z(k-1) = \begin{cases} \lambda_i TP_z(k-1), & k \geq 1, \\ 0, & k < 1, \end{cases}$$

$$M_j(k)P_z(k) = \mu_j TP_z(k-1) = \begin{cases} \mu_j TP_z(k-1), & k \geq 1, \\ 0, & k < 1. \end{cases} \qquad (5.17)$$

Taking into account the transformations of the products in the image field (5.17), the system of spectral Eq. (5.16) will look like the following:

$$\begin{cases} P_{S0}(k) = \prod_{i=1}^{4} P_{Si}(k); \\ P_{F0}(k) = 1 - \prod_{i=1}^{4}(1 - P_{Fi}(k)); \\ P_{Si}(k) = 1 - P_{Fi}(k) - P_{SFi}(k); \\ P_{SFi}(k+1) = \dfrac{T^2}{k+1}[-(\lambda_i + \mu_i)P_{SFi}(k-1) + \lambda_i P_{Si}(k-1) + \mu_i P_{Fi}(k-1)]; \\ P_{Fi}(k+1) = \dfrac{T^2}{k+1}[-\mu_i P_{Fi}(k-1) + \lambda_i P_{SFi}(k-1)], \quad i = 1..4. \end{cases} \qquad (5.18)$$

Using the spectral model of operation of the system (5.18) and the initial conditions (5.6), we determine the probability discrete $P_{Fi}(k)$, and as $k := 0, 1, 2\ldots$, a result we obtain is as follows:

$$P_{Fi}(0) = P_{Fi}(1) = P_{Fi}(2) = 0; \qquad (5.19)$$

$$P_{Fi}(3) = \frac{T^4}{3}\lambda_i^2; \qquad (5.20)$$

$$P_{Fi}(4) = \frac{T^4}{8}(\lambda_i^2 - T^2\lambda_i^3); \qquad (5.21)$$

$$P_{Fi}(5) = \frac{T^4}{15}(-2T^2\mu_i\lambda_i^2 - 2T^2\lambda_i^3 + \lambda_i^2); \qquad (5.22)$$

$$P_{Fi}(6) = \frac{T^4}{24}(-T^2\mu_i\lambda_i^2 + T^4\mu_i\lambda_i^3 + T^4\lambda_i^4 - T^2\lambda_i^3 + \lambda_i^2). \qquad (5.23)$$

According to the second equation of the system (5.18) and expressions (5.19)–(5.23), we obtain the corresponding discrete probability of server failure:

$$P_{F0}(0) = P_{F0}(1) = P_{F0}(2) = 0; \qquad (5.24)$$

$$P_{F0}(3) = 1 - \prod_{i=1}^{4}\left(1 - \frac{T^4}{3}\lambda_i^2\right); \qquad (5.25)$$

$$P_{F0}(4) = 1 - \prod_{i=1}^{4}\left(1 - \frac{T^4}{8}(\lambda_i^2 - T^2\lambda_i^3)\right); \qquad (5.26)$$

5 Green Technologies of Information Protection in Computer Networks... 85

$$P_{F0}(5) = 1 - \prod_{i=1}^{4}\left(1 - \frac{T^4}{15}\left(-2T^2\mu_i\lambda_i^2 - 2T^2\lambda_i^3 + \lambda_i^2\right)\right); \qquad (5.27)$$

$$P_{F0}(6) = 1 - \prod_{i=1}^{4}\left(1 - \frac{T^4}{24}\left(-T^2\mu_i\lambda_i^2 + T^4\mu_i\lambda_i^3 + T^4\lambda_i^4 - T^2\lambda_i^3 + \lambda_i^2\right)\right). \qquad (5.28)$$

5.6 Determining the Main Optimal Strategies for Protecting Information in Cyberattacks

To obtain the price of the game in the image area, apply the differential transformation (5.15) to the expression of the price of the game in the time area (Korobiichuk et al. 2018, 2019; Demin 2017) (5.13), and then we obtain the following:

$$I = \sum_{k=0}^{\infty} \frac{P_{F0}(k)}{k+1}. \qquad (5.29)$$

Taking into account (5.29), accordingly, in the field of images, we will consider the maximum criterion (5.11), which we formalize in the following form:

$$\min_{\mu_{jk}\in E_\mu} \max_{\lambda_{ik}\in E_\lambda} I = \max_{\lambda_{ik}\in E_\lambda} \min_{\mu_{jk}\in E_\mu} I = I^*\left(t, P_{F0}^{opt}(k), \lambda_{ik}^{opt}, \mu_{jk}^{opt}\right). \qquad (5.30)$$

Find the price of the game by substituting discrete (5.24)–(5.28) on (5.29), we get the following:

$$I \approx \sum_{k=0}^{6} \frac{P_{F0}(k)}{k+1} = 1 + \frac{1}{4}\left(1 - \prod_{i=1}^{4}\left(1 - \frac{T^4}{3}\lambda_i^2\right)\right) + \frac{1}{5}$$
$$\times \left(1 - \prod_{i=1}^{4}\left(1 - \frac{T^4}{8}\left(\lambda_i^2 - T^2\lambda_i^3\right)\right)\right) +$$
$$+ \frac{1}{6}\left(1 - \prod_{i=1}^{4}\left(1 - \frac{T^4}{15}\left(-2T^2\mu_i\lambda_i^2 - 2T^2\lambda_i^3 + \lambda_i^2\right)\right)\right) +$$
$$+ \frac{1}{7}\left(1 - \prod_{i=1}^{4}\left(1 - \frac{T^4}{24}\left(-T^2\mu_i\lambda_i^2 + T^4\mu_i\lambda_i^3 + T^4\lambda_i^4 - T^2\lambda_i^3 + \lambda_i^2\right)\right)\right). \qquad (5.31)$$

Find the extremums of the function (5.3–5.31) for which we solve the system of the following algebraic equations:

$$\begin{cases} \dfrac{\partial I}{\partial \lambda_i} : \left[\dfrac{T^4}{6}\lambda_i\right]\left(1 - \prod_{j=1, j\neq i}^4 \left(1 - \dfrac{T^4}{3}\lambda_j^2\right)\right) + \left[\dfrac{T^4}{40}(2\lambda_i - 3T^2\lambda_i^2)\right] \\ \left(1 - \prod_{j=1, j\neq i}^4 \left(1 - \dfrac{T^4}{8}(\lambda_j^2 - T^2\lambda_j^3)\right)\right) + \left[\dfrac{T^4}{45}(-2T^2\mu_i\lambda_i - 3T^2\lambda_i^2 + \lambda_i)\right] \\ + \left(1 - \prod_{j=1, j\neq i}^4 \left(1 - \dfrac{T^4}{15}(-2T^2\mu_j\lambda_j^2 - 2T^2\lambda_j^3 + \lambda_j^2)\right)\right) + \\ + \left[\dfrac{T^4}{168}(-2T^2\mu_i\lambda_i + 3T^4\mu_i\lambda_i^2 + 4T^4\lambda_i^3 - 3T^2\lambda_i^2 + 2\lambda_i)\right] \times \\ \times \left(1 - \prod_{j=1, j\neq i}^4 \left(1 - \dfrac{T^4}{24}(-T^2\mu_j\lambda_j^2 + T^4\mu_j\lambda_j^3 + T^4\lambda_j^4 - T^2\lambda_j^3 + \lambda_j^2)\right)\right) = 0; \\ \dfrac{\partial I}{\partial \mu_i} : \left[-\dfrac{T^6}{45}\lambda_i^2\right]\left(1 - \prod_{j=1, j\neq i}^4 \left(1 - \dfrac{T^4}{15}(-2T^2\mu_j\lambda_j^2 - 2T^2\lambda_j^3 + \lambda_j^2)\right)\right) + \\ + \left[\dfrac{T^6}{168}\lambda_i^2(T^2\lambda_i - 1)\right] \\ \left(1 - \prod_{j=1, j\neq i}^4 \left(1 - \dfrac{T^4}{24}(-T^2\mu_j\lambda_j^2 + T^4\mu_j\lambda_j^3 + T^4\lambda_j^4 - T^2\lambda_j^3 + \lambda_j^2)\right)\right) = 0. \end{cases}$$

$$(5.32)$$

Taking into account the condition of independence of cyberattacks, protection strategies for each of the information tasks of the subway distance server can be found by assuming the absence of conflicts over other functions. Thus, we obtain the following:

$$\begin{cases} \dfrac{\partial I}{\partial \lambda_i} : \dfrac{T^4}{6}\lambda_i + \dfrac{T^4}{40}(2\lambda_i - 3T^2\lambda_i^2) + \dfrac{T^4}{45}(-2T^2\mu_i\lambda_i - 3T^2\lambda_i^2 + \lambda_i) + \\ + \dfrac{T^4}{168}(-2T^2\mu_i\lambda_i + 3T^4\mu_i\lambda_i^2 + 4T^4\lambda_i^3 - 3T^2\lambda_i^2 + 2\lambda_i) = 0; \\ \dfrac{\partial I}{\partial \mu_i} : -\dfrac{T^6}{45}\lambda_i^2 + \dfrac{T^6}{168}\lambda_i^2(T^2\lambda_i - 1) = 0. \end{cases} \qquad (5.33)$$

After performing the appropriate transformations of the system (5.33), we obtain expressions for the optimal strategies of players in the field of images:

$$\begin{cases} \mu_i = -\dfrac{60T^4\lambda_i^2 - 402T^2\lambda_i + 602}{45T^4\lambda_i - 142T^2}; & \mu_i \approx 16.35\dfrac{1}{T^2}; \\ \lambda_i = \dfrac{213}{45}\dfrac{1}{T^2}; & \lambda_i \approx 4.73\dfrac{1}{T^2}. \end{cases} \qquad (5.34)$$

5 Green Technologies of Information Protection in Computer Networks...

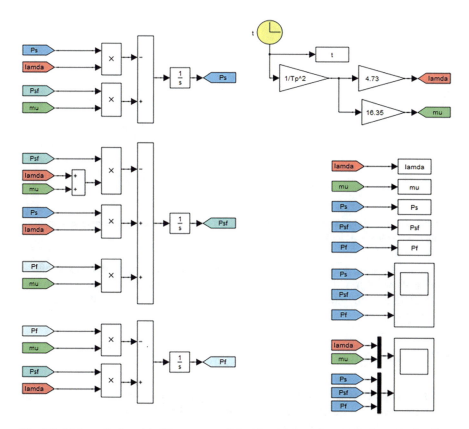

Fig. 5.2 Mathematical model of the process of attacking cyberattacks on a single server function

After moving to the time domain according to (5.15), we obtain expressions for optimal defense strategies and cyberattacks:

$$\begin{cases} \mu_i(t) \approx 16.35 \dfrac{t}{T^2}; \\ \lambda_i(t) \approx 4.73 \dfrac{t}{T^2}. \end{cases} \quad (5.35)$$

On the basis of (5.3) and (5.6) using (5.35), we build a mathematical model of the process of cyberattack on a separate function of the server of the power supply system in the MATLAB environment, namely, in the Simulink part (Fig. 5.2).

This graph (Fig. 5.3) presents the results of mathematical modeling of the process of changing the probabilities of states of the information system in the process of cyberattack (Voronko 2013).

The given graphs illustrate the dynamics of the information conflict in the considered power supply system in the conditions of attack on the selected function

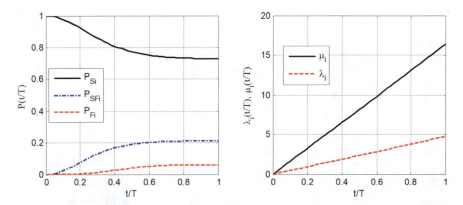

Fig. 5.3 Graphs of processes of probabilities of states of some function of the server at implementation of cyberattack with the use of optimum strategies

of the multitasking server of the subway distance. The simulation of the cyberattack process was performed to assess the integrated security indicators of the server based on (5.2), (5.3), and (5.6), using the optimal strategies (5.35) for each of its functions.

5.7 Conclusions

1. The computer system of monitoring and diagnostics of parameters of modes of electric power systems of the underground is analyzed, namely, theoretical bases and methods of information protection.
2. A large number of tasks of the main server of the power supply system of the subway distance, which require information protection, can be achieved with the help of models and methods of information protection of the components of the power supply network.
3. Mathematical models of the process of cyberattack on the multitasking server of the computer system for monitoring and diagnostics of energy facilities of the subway allow to obtain optimal strategies for protecting information in cyberattacks. At the same time, methods of differential game modeling of cyberattack and information protection processes were used, for which increased requirements are set for reliability and information security.
4. Computer simulation is carried out, as a result, when using the obtained optimal strategies, and the guaranteed probability level of protection of a separate function of the subway power supply server in the conditions of a concentrated cyberattack at the end of the conditional review period is 0.73. At the same time, the probability of server failure for some functionality is 0.06. If the players involved in the information conflict deviate from the optimal strategies, the values of the probabilities of states will change accordingly, but the value of the fee will not correspond to the minimum.

References

Borankulova GS, Tungatarova AT (2019) Methods and means of information network protection. ISJ Theor Appl Sci 04(72):75–78

Demin DA (2017) Synthesis of optimal control over inertial technological processes based on a multialternative parametric description of the final state. East Eur J Enterp Technol 3/4(87):51–63

Fomin O, Lovska A, Kulbovskyi I et al (2019) Determining the dynamic loading on a semi-wagon when fixing it with a viscous coupling to a ferry deck. East-Eur J Enterp Technol 2(7(98)):6–12

Holub H, Soloviova O (2019) Research of information protection methods in computer system for monitoring and diagnostics of electric supply of railways. In: Technology transfer: fundamental principles and innovative technical solutions, pp 9–11. https://doi.org/10.21303/2585-6847.2019.001028

Holub H, Skok KI et al (2020) System model of information flows in networks of the electric supply system in transport infrastructure projects. Transport Means:132–135

Hryshchuk RV (2009) Metod dyferentsial'no-ihrovoho R-modelyuvannya protsesiv napadu na informatsiyu. Informatsiyna bezpeka 2(2):128–132

Hryshchuk RV (2010) Teoretychni osnovy modelyuvannya protsesiv napadu na informatsiyu metodamy teoriy dyferentsial'nykh ihor ta dyferentsial'nykh peretvoren'. Ruta, Zhytomyr

Ihnatov V, Huziy M (2005) Dynamika informatsiynykh konfliktiv v intelektual'nykh systemakh. Problemy informatyzatsiyi ta upravlinnya 15:88–92

Khoroshko VA, Morzhov SV (2004) Application of petry networks to parallel process modeling. J Autom Inf Sci 36(4):12–17

Korobiichuk I, Hryshchuk R, Mamarev V et al (2018) Cyberattack classificator verification. Adv Intell Syst Comput 635:402–411

Korobiichuk I, Hryshchuk R, Horoshko V et al (2019) Microprocessor means for technical diagnostics of complex systems. CEUR Workshop Proc 2353:1020–1029

Kulbovskyi I, Bakalinsky O, Sorochynska O et al (2019) Implementation of innovative technology for evaluating high-speed rail passenger transportation. Eureka Phys Eng 6:63–72

Lenkov S, Peregudov D, Khoroshko V (2008) Metody i sredstva zashchity informatsii: monografiya [v 2-kh t.]: T.1: Informatsionnaya bezopasnost'. Ariy, Kiїv

Nguyen T, Wright M, Wellman M et al (2018) Multistage attack graph security games: heuristic strategies, with empirical game-theoretic analysis. Secur Commun Netw 2018:1–28

Pukhov GY (1986) Differentsial'nyye preobrazovaniya i matematicheskoye modelirovaniye fizicheskikh protsessov: monografiya. Naukova dumka, Kiїv, p 160 c

Song J, Lee Y, Kim K et al (2015) Automated verification methodology of security events based on heuristic analysis. Int J Distrib Sens Netw. https://doi.org/10.1155/2015/817918

Stasiuk O, Honcharova L, Maksymchuk V (2012) Metody orhanizatsii kompiuternoi merezhi monitorynhu parametriv rezhymiv system elektropostachannia. Informatsiino-keruiuchi systemy na zaliznychnomu transporti, naukovo-tekhnichnyi zhurnal 2:35–40

Vorobiev S, Petrenko I, Kovaleva I et al (2017) Analysis of computer security incidents using fuzzy logic. In: Proceedings of the 20th IEEE international conference on soft computing and measurements, pp 369–371

Vorob'yov AA, Kulikov GV, Nepomnyashchikh AV (2007) Otsenivaniye zashchishchonnosti avtomatizirovannykh sistem na osnove metodov teorii igr. Informatsionnyye tekhnologii. Novyye tekhnologii, Moskva, p 24

Voronko IO (2013) Dyferentsial'no – ihrova model' nadiynosti mikroprotsesornykh system monitorynhu tyahovykh elektrychnykh merezh zaliznyts'. Informatsiyno-keruyuchi systemy na zaliznychnomu transporti 5:8–15

Chapter 6
Multistage Drying in Fluidized Bed: Ways of Eco-friendly Application and Marketing Tools for Promotion

Nadiia Artyukhova, Tetiana Vasylieva, Serhiy Lyeonov, Jan Krmela, Oleksandr Shandyba, and Olena Melnyk

6.1 Introduction

Hydromechanical, heat, and mass transfer processes occurring in the "gas"-"solid" system are most common in chemical production. They include separation (classification) of granular (dispersed) materials, heat treatment in a stream of hot or cold heat transfer agents, dehydration, etc. In two-phase systems, "gas"-"solid," the dispersed phase quite often is a polydisperse system, the components of which (due to different force effects on particles of different sizes) can move differently in a continuous flow (gas phase). The change in the mass of a particle due to the drying in a stream of a hot heat transfer agent's motion can also complicate the motion process.

The polydisperse system in the device (the particles of which can change their mass in time) leads to the fact that small particles and dust can be carried away from the device with waste heat transfer agent influenced by the gas flow (which predominantly has a constant velocity in the classical convective drying device). This problem defines the relevance of searching for ways to reduce the concentration of solids in the gas flow leaving the dryer's workspace.

N. Artyukhova (✉) · T. Vasylieva · S. Lyeonov
Sumy State University, Sumy, Ukraine
e-mail: n.artyukhova@pohnp.sumdu.edu.ua; tavasilyeva@biem.sumdu.edu.ua;
s.lieonov@uabs.sumdu.edu.ua

J. Krmela
Alexander Dubcek University of Trencin, Puchov, Slovakia
e-mail: jan.krmela@fpt.tnuni.sk

O. Shandyba · O. Melnyk
Sumy National Agrarian University, Sumy, Ukraine
e-mail: olena.melnyk@snau.edu.ua

© The Author(s), under exclusive license to Springer Nature Switzerland AG 2023
S. Boichenko et al. (eds.), *Sustainable Transport and Environmental Safety in Aviation*, Sustainable Aviation, https://doi.org/10.1007/978-3-031-34350-6_6

One of the most successful hydrodynamic systems applied to convective drying is the "fluidized bed" system. Its advantages are noted in a couple of works (Kwauk 1992; Gidaspow 1994; Yang 2003; Gibilaro 2001) and for specific processes (Horio 2013; Kuo et al. 2021; Zhong et al. 2008). Scientists also studied environmental safety during the operation of devices with a fluidized bed (Mazza et al. 2016; Koshkarev et al. 2017) and the constructive improvement of equipment to reduce harmful emissions into the atmosphere (Kosowska-Golachowska et al. 2019) and water (Melnyk et al. 2020) in different ways (Melnyk 2016).

Methods for forming a fluidized bed of dispersed particles are divided according to the implementation way (Stahl 2004; Parikh 2009; Muralidhar et al. 2016), flow motion organization (Stahl 2010; Litster and Ennis 2004), instrumentation (Solanki et al. 2010; Saikh 2013), the use of additional stages (fundamental aspects (Artyukhova 2020)), computer simulation (Artyukhova and Krmela 2019), and constructive solutions (Prokopov et al. 2014) with various hydrodynamic and thermodynamic conditions (Obodiak et al. 2020), which differ in their advantages, disadvantages, and application field (Patel et al. 2011).

At the dryers' design stage (before the physical experiment), it is necessary to make an optimization calculation without additional costs for physical models. In this case, simulation (computer) modeling enables a rational equipment design during the theoretical description. In this regard, computer modeling (as described in the application to hydromechanical (Tahvildari et al. 2016), heat (Wang et al. 2015), mass transfer (Chen et al. 2007) (in particular, granulation with the description of modeling results (Artyukhov et al. 2017a, b)), simulation of the vortex fluidized bed functioning (Artyukhov et al. 2020a, b), convective heat transfer (Artyukhov et al. 2020a, b), and reaction processes (Trojan 2015; Liu et al. 2012) are an urgent need for the development of an engineering method to calculate environmentally friendly drying equipment.

The drying process in the multistage device with vertical sectioning of its workspace should be distinguished among the various convective drying methods (Krmela et al. 2020). This dryer has a reliable theoretical description (Artyukhov et al. 2017a, b), confirmed by experimental studies (Artyukhov and Artyukhova 2018), preliminary computer modeling results (Artyukhov and Artyukhova 2019), and successful implementation in production (Artyukhov et al. 2017a, b).

Developing comprehensive measures to introduce new solutions into production after fulfilling the dryer's environmental safety conditions is necessary. Therefore, the technology readiness level should be evaluated, and marketing tools for promoting the development to the market should be offered (commercialization success of innovations (Andrade and Loureiro, 2020), sustainable business models (Lipkova and Braga, 2016), managing enterprise sustainable development (Bilan et al, 2020)). It is also essential to follow the goals of sustainable development (Kasych and Vochozka 2017) and innovative tools for creating new technologies (Kobushko et al. 2017; Spremberg et al. 2017).

6 Multistage Drying in Fluidized Bed: Ways of Eco-friendly Application... 93

Thus, the goals of this work are as follows:

1. To create and justify the environmental safety of new shelf dryers' designs.
2. To perform computer modeling of the hydrodynamic conditions of the flow motion in the dryer to define the optimal technological conditions and the device's design that meet the environmental safety requirements.
3. To create a model for the new dryers' designs at the market, assessment of the technological readiness of investigations, and the prospects for their further improvement.

6.2 Computer Modeling as a Tool for the Optimization Design of Environmentally Friendly Dryers

Evaluation of the influence factors of the dryer's design on the nature of the gas flow motion is possible at the simulation (computer) modeling stage as a kind of optimization experiment. During this experiment, the shelf length, its tilt angle, and the perforation degree of the shelf (the free section size) are selected, considering the peculiarities of the hydrodynamic system "fluidized bed." The stability of the fluidized bed is determined by the range of gas flow velocities between the first (beginning of fluidization) and the second (ablation of particles from the fluidized bed) critical velocities (Artyukhova 2020). Determining the fluidized bed range in devices with a constant height cross section does not complicate the monodisperse system processing. Fluidization of polydisperse systems presents some difficulties due to the need to determine the range of critical velocities. It increases the probability of the small particles' ablation from the device and increases harmful emissions.

The gas flow motion modeling with a dryer's variable cross-sectional area is a more difficult task (in terms of predicting the amount of solid particle emissions into the atmosphere with outgoing gases). The calculated model of the dryer (Fig. 6.1) enables us to see that the shelf contact installation changes the section of the device; in this case, several peculiar zones are formed in the device:

- Overlayer space.
- Separation zone.
- Outloading gap.

For the analysis of turbulent flows, the Reynolds equation and the continuity of the flow are used (Artyukhov and Sklabinskiy 2017):

$$\frac{\partial}{\partial \tau}\left(\rho_{cf}\overline{u_{cfi}}\right) + \frac{\partial}{\partial x_j}\left(\rho_{cf}\overline{u_{cfi}u_{cfj}}\right) + \frac{\partial}{\partial x_j}\left(\rho_{cf}\overline{u'_{cfi}u'_{cfj}}\right) = -\frac{\partial p}{\partial x_i} + \frac{\partial}{\partial x_j}\left[\mu_{cf}\left(\frac{\partial \overline{u_{cfi}}}{\partial x_j} + \frac{\partial \overline{u_{cfj}}}{\partial x_i}\right)\right] + f_i, \quad (6.1)$$

$$\frac{\partial \rho_{cf}}{\partial \tau} + \frac{\partial}{\partial x_j}\left(\rho_{cf}u_{cfj}\right) = 0, \quad (6.2)$$

where $\overline{u_{cf1}}, \overline{u_{cf2}}, \overline{u_{cf3}}$ is the time-averaged velocities of carrier phase and $\overline{u'_{cf1}}, \overline{u'_{cf2}}$ is the pulsation component of velocities of carrier phase.

Fig. 6.1 The model of the dryer and construction of a computational grid for modeling

In Eqs. 6.1 and 6.2, the simplified equations are used, $i, j = 1 \ldots 3$, and the summing up to over the same indices is assumed, x_1, x_2, x_3 – coordinate axes, τ – time. The f_i term expresses the action of mass forces.

Each of the shelves can have different design and installation features. The number of shelves and their design are chosen, providing the residence time of granules for drying in the device. Therefore, selecting the optimal gas flow rate on a physical model is materially unprofitable, given many zones with different cross-sectional areas in the device. A physical experiment can be carried out only after optimizing the dryer's design at the computer modeling stage.

The results of modeling the gas flow process in the dryer are shown in Figs. 6.2 and 6.3.

Analysis of the computer modeling results allows the following:

- To establish zones of the increased and reduced intensity of the gas flow and predict (depending on the fractional composition of dispersed material) the amount of small particle ablation from the device.
- To define the vortex formation zones, leading to a violation of the gas flow structure and the removal of particles from the device.

6 Multistage Drying in Fluidized Bed: Ways of Eco-friendly Application... 95

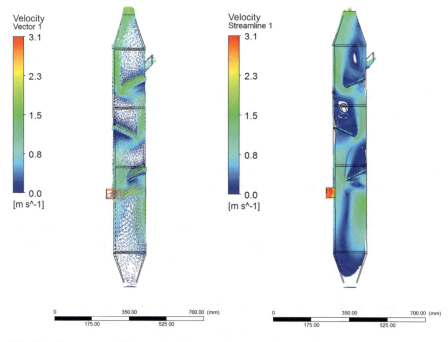

Fig. 6.2 Computer modeling results of the gas flow motion in a shelf dryer

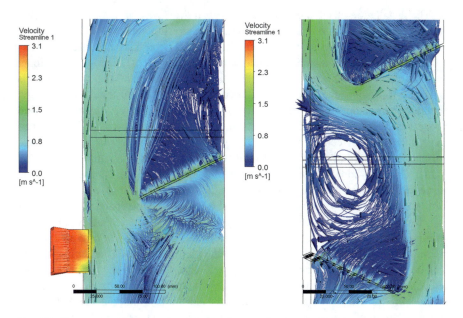

Fig. 6.3 Computer modeling results of the gas flow motion in a shelf dryer

- To compare the calculated critical velocities of the gas flow with the velocities in different zones of the device and draw conclusions about the motion of particles of different diameters.
- To study the distribution of velocities in the dryer's separation space and to propose the optimal design of that zone to reduce dust carried away from the device.

The computer modeling results form the basis for developing new designs for shelf contacts, some of which are given below.

6.3 Ways to Improve the Environmental Safety of Multistage Fluidized Bed Dryers

The developed constructive solutions are based on improving the dispersed material drying device by changing the design of contact shelves. It increases the uniformity of the drying agent's contact with the dispersed material flow. It improves flow motion hydrodynamics to increase the contact time between dispersed material and a drying agent.

The following improvements to the dryer's construction are proposed:

- Installation of shelf contacts with different outloading gap up the dryer's height.
- Use of shelf contact with variable perforation.
- Execution of a prefabricated shelf contact, each part of which has a different tilt angle to the horizon.

The installation of inclined contact shelves in the device with a gap of different heights increases the drying efficiency during the moisture removal from the surface layer of the dispersed material and from the material depth after heating.

When the dispersed material contacts with the drying agent (heating) on the upper inclined contact shelf where the gap has the maximum value, the shelf length is minimal. It provides the minimum required contact time of phases and full heating without overheating the dispersed material, which can negatively affect the drying process of thermolabile materials. Simultaneously, it removes the small fraction, i.e., the upper inclined contact shelf serves as a separator, positively impacting the formation of a weighted layer on subsequent contact shelves by aligning the porosity value. On the middle-inclined contact shelf, where the gap is smaller when the dispersed material comes into contact with the drying agent (removal of moisture from the surface layer), the residence time of the dispersed material and contact with the drying agent grows due to increasing the length of the contact shelf, which promotes intensive removal of unbound moisture. The residence time of the dispersed material on the shelf corresponds to the required value of the drying time in this period (the period of constant drying velocity). On the lower inclined contact shelf, when the dispersed material comes into contact with the drying agent (removal of moisture from the material depth), where the gap value is minimal, the shelf length is maximum. It effectively removes bound moisture from the material depth and the maximum required contact time of the phases in this period (decline of the drying velocity).

Perforation of the inclined contact shelf with holes of different diameters allows to create a hydrodynamic situation when the drying agent's velocity motion epure is aligned on the shelf length, and its action on all length remains constant. It causes the compensation of inertia effect on the dispersed material and rolling on an inclined surface, braking the dispersed material on an inclined contact shelf, its uniform motion in the suspended layer, and long-term contact with the drying agent.

Due to the constant flow rate of the drying agent in each of the cross sections of the device, the gap between the end of the inclined contact shelf and the wall, different free cross section of the drying agent in individual parts of the device, and the installation of inclined contact shelves will reduce the drying agent's velocity along the length of the inclined contact and in the gap between the end of the inclined contact shelf and the wall. It helps to increase the contact uniformity of the drying agent with the dispersed material.

The mentioned construction of inclined contact shelves helps reduce the vortex formation's intensity due to the compensation of forces forming a vortex when bending the end of the inclined contact shelf, increasing the drying agent's ascending flow force.

The installation of inclined contact shelves in the device's workspace with a certain perforation helps to increase the drying efficiency during the removal of moisture from the surface layer of the dispersed material and from the material depth after heating.

Execution of the inclined perforated contact shelf formed with a changing tilt angle to the horizon allows creating a hydrodynamic situation when the drying agent's alignment along the length of the inclined perforated contact shelf, its action along its entire length remains constant. It leads to the compensation of the inertial force effect on the dispersed material and rolling on an inclined surface, breaking the dispersed material on an inclined perforated contact shelf, its uniform motion in the weighted layer, and long-term contact with the drying agent.

The construction of inclined perforated contact shelves helps to reduce the vortex process intensity due to the compensation of forces forming a vortex when bending the end of the inclined contact shelf, increasing the force of the drying agent's ascending flow.

The use of inclined perforated contact shelves of the above structures creates a hydrodynamic situation at each dryer stage, where the drying agent's velocity epure is aligned along the shelf length. Its action remains constant in all areas of the shelf. It causes the compensation of the action on the dispersed material of inertial forces and rolling on an inclined surface, braking the dispersed material on an inclined perforated contact shelf, its uniform motion in the weighted layer, and long-term contact with the drying agent.

By developing mechanisms for controlling the motion of particles on the shelf and predicting the particle distribution law in the device (while obtaining a reliable and uniform distribution of the gas flow), solid particle emissions with outgoing gas can be significantly reduced. The comparative analysis of the solid particle

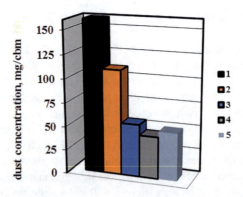

Fig. 6.4 The average content of ammonium nitrate dust in the gases outgone from the dryer: 1 – fluidized bed dryer; 2 – dryer with shelves with a constant length, tilt angle, and perforation; 3 – shelf with a different gap up the dryer's height; 4 – sectioned shelf with variable perforation of sections; 5 – sectioned shelf with constant perforation of sections and variable tilt angle

concentration in the devices with different designs of shelves (in comparison with classical fluidized bed devices) is presented in Fig. 6.4. The particles to be dried are granules of porous ammonium nitrate.

6.4 Introduction of Multistage Dryers into Production: Marketing Tools

Scenario (roadmap of technological development) "environmentally friendly multistage shelf dryer" is a time-defined set of measures to introduce the technology, declared by the investigator (designers and technologists) and described by the executor (marketing service) based on the algorithm, on the market.

The roadmap measures aim to create a competitive drying technology in the investigator's device under limited investment opportunities at the national level.

The purpose of the roadmap is to increase the efficiency of using technological equipment for the drying process by transitioning to a new multistage scheme with economically sound technologies that will further reduce pollutants' emissions into the atmosphere.

The roadmap "environmentally friendly multistage shelf dryer" is formed by the following way:

Stage 1

A comprehensive analysis of the technological readiness of the unit and its commercialization level are given in Fig. 6.4 (tool – NYSERDA 2020).

The results of the data analysis define further measures (stages) to develop the technological level of equipment. Some of them can be carried out in parallel. Still, it is obligatory in some cases (e.g., forming a team and transitioning from technology to industrial design of the unit) (Fig. 6.5).

Technology & Commercialization Readiness Level Calculator

Profile	
Company/Organization Name:	Sumy State University
Proposal Title:	Multistage shelf dryer
Product/Innovation Description:	Manufacturing of porous ammonium nitrate

Technology Readiness Level: 7

Commercialization Readiness Level: 3

Category	Answer
Technology	Initial testing of integrated product/system has been completed in a laboratory environment
Product Development	Demonstration of a full scale product/system prototype has been completed in the intended application(s)
Product Definition/Design	The product/system has been scaled from laboratory to pilot scale and issues that may affect achieving full scale have been identified
Competitive Landscape	Comprehensive market research to prove the product/system commercial feasibility has been completed and intermediate understanding of competitive products/systems has been demonstrated
Team	Balanced team with technical and business development/commercialization experience running the company with assistance from outside advisors/mentors
Go-To-Market	Market and customer/partner needs and how those translate to product requirements have been defined, and initial relationships have been developed with key stakeholders across the value chain
Manufacturing/Supply Chain	Products/systems have been pilot manufactured and sold to initial customers

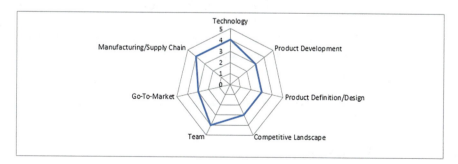

Fig. 6.5 Comprehensive analysis of the technological readiness level of the unit and its commercialization level

1. Technology

   - Continuing the applied research and defining the practical application.
   - Completing preliminary tests of technological components and establishing technical expediency conditions of the further introduction in the laboratory.
   - Evaluating the effectiveness of the technology using a multifactor experiment (technological parameters of the unit + optimal design of shelf contacts).

2. Level of Product Development (Completion of Technology)

   - Assessment of product compliance with market needs.
   - Product testing in production conditions.
   - Substantiating the product capacity at work in production conditions.
   - Creating an evidence base regarding the production efficiency in the operating conditions and environment.

3. Product Design and Scaling (Creation of an Industrial Design of the Unit)

   - Analysis and comparison of data on the theoretical description and experimental researches of the drying process in the device. Formation of the comprehensive model for calculating the installation.
   - Identifying the client's specific needs (technical task from the business representative) to create the sample of the drying unit.
   - Transferring the laboratory unit to the industrial one, identifying the problems that may affect the achievement of full scale.
   - Development of a comprehensive model (commercial offer) for potential customers.
   - Optimization of the final design of the industrial unit for the potential customer's needs.

4. Analysis of the Competitive Environment (Market)

   - Completing the primary market research to confirm the commercial feasibility of further technological development.
   - Comprehensive market research to confirm the commercial feasibility of further technological development.
   - Analysis of the existing unique features and advantages of technology in comparison with competitive technologies.
   - Creation of the competitive landscape map (complete understanding of the market structure and target application of the developed technology).

5. A Team of the Project

   - Creation of a balanced team with technical and business development/commercialization.
   - Amendment of the team with new members (functions: sales, marketing, customer service, etc.)
   - Transition to the market.

- Survey of potential clients and partners (possibly donors) of the project identifying weak points of technology and creating (further – improvement) business model of technology sale.
- Formation of the network of clients and partners (possibly donors).
- Direct conclusion of technology sales contracts.

6. Establishing Relationships in the Market

    - Realization of the clients' network potential.
    - Establishing relations with suppliers of intermediate services, partners, suppliers of materials (components) for the drying unit.
    - Technology (unit) sale on a single number.
    - Full-scale production and formation of a supply chain (technology or unit).

The above measures are structurally combined into one scheme, a prototype of roadmaps for technological development of the project "environmentally friendly multistage shelf dryer" (Fig. 6.6).

The algorithm for investigating the roadmap in the project "environmentally friendly multistage shelf dryer" is presented in Fig. 6.7. It is based on own research (Artyukhov et al. 2016) and considers the data from the literature sources describing the roadmap formation (Phaal et al. 2004; Münch et al. 2019).

According to the roadmaps' classification (Phaal et al. 2004), the observed roadmap corresponds to level 2 "expert opinion."

The roadmap includes four stages. During their implementation, the developer must perform the following:

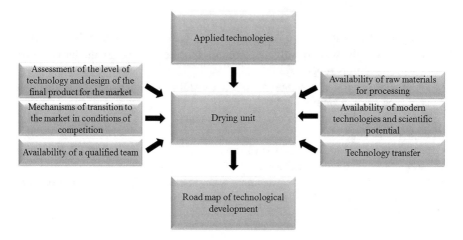

Fig. 6.6 The structure of the technological development roadmap of the project "environmentally friendly multistage shelf dryer"

Fig. 6.7 Algorithm to investigate a roadmap of the technological development of the project "environmentally friendly multistage shelf dryer"

- Study the national strategy (programs) of industry development from general challenges to the ways of solving them in a specific area, where the introduction of a drying unit is proposed.
- Analyze in detail the existing infrastructure of that sector in which the drying unit is proposed to introduce.
- Review current drying technologies to find a niche for the technology offered for implementation.
- Make the way from the creation of research and industrial unit to the industrial model of the chemical enterprise.
- Support each stage of the roadmap with marketing research and economic calculations of the claimed technology attractiveness.

6.5 Conclusions

This paper presents an algorithm for assessing the technological level of development readiness for implementation on the market, which consists of the following stages:

- Modeling of the main processes that are implemented in the dryer.
- Selection of the optimal design of the drying apparatus based on the simulation results.
- An experimental study of the effectiveness of the chosen constructive solution.
- The creation of a roadmap for the implementation of the development.
- Choosing the best way to carry out each stage of the development implementation.

At each stage, the optimal characteristics of the drying equipment are determined, which can be competitive compared with similar equipment using other design and technological solutions. New designs of shelf units allow controlling the motion of dispersed particles in the device. At the same time, it is possible to predict the behavior of dust and small particles and ensure their reliable trapping from outgoing gases. An additional tool for reducing the harmful emissions of the solid phase (at the preliminary calculation) is the computer modeling of the dryer's hydrodynamic conditions.

The obtained results have prospects for further implementation due to the analysis of the dryer's technology readiness level and the roadmap development for equipment to enter the domestic market.

Acknowledgments The authors thank researchers at the Marketing Department, Sumy State University, Department of Numerical Methods and Computational Modeling, and Department of Materials Technologies and Environment, Alexander Dubcek University of Trenčín, for their valuable comments during the article preparation.

This research work had been supported by the Ministry of Science and Education of Ukraine under the project "Creation of New Designs of Equipment for Porous Ammonium Nitrate Obtaining: Energy-Efficient and Environmentally Friendly Solutions," by the Cultural and Educational Grant Agency of the Slovak Republic (KEGA) under the project "Simulations of Basic and Specific Experiments of Polymers and Composites Based on Experimental Data in Order to Create a Virtual Computational-Experimental Laboratory for Mechanical Testing" (No. KEGA 003TnUAD-4/2022) and under the project "Advancement and Support of R&D for 'Centre for Diagnostics and Quality Testing of Materials' in the Domains of the RIS3 SK Specialization," code NFP313011W442.

References

Andrade HS, Loureiro GA (2020) Comparative analysis of strategic planning based on a systems engineering approach. Bus Ethics Leadersh 4(2):86–95

Artyukhov A, Artyukhova N (2018) Utilization of dust and ammonia from exhaust gases: new solutions for dryers with different types of fluidized bed. J Environ Health Sci Eng 16(2):193–204

Artyukhov AE, Artyukhova NO (2019) Technology and the main technological equipment of the process to obtain NH_4NO_3 with nanoporous structure. Springer Proc Phys 221:585–594

Artyukhov AE, Sklabinskiy VI (2017) Application of vortex three-phase separators for improving the reliability of pump and compressor stations of hydrocarbon processing plants. IOP Conf Ser Mater Sci Eng 233(1):012014

Artyukhov AY, Omelyanenko VA, Artyukhova NO (2016) Strategic framework and methodical bases of technological package development management. Mark Manag Innov 3:170–179

Artyukhov AE, Obodiak VK, Boiko PG et al (2017a) Computer modeling of hydrodynamic and heat-mass transfer processes in the vortex type granulation devices. CEUR workshop proceedings 1844, pp 33–47

Artyukhov A, Artyukhova N, Ivaniia A (2017b) Progressive equipment for generation of the porous ammonium nitrate with 3D nanostructure. In: Proceedings of the 2017 IEEE 7th international conference on nanomaterials: applications and properties, NAP 2017, 2017-January, 03NE06

Artyukhov A, Artyukhova N, Krmela J et al (2020a) Granulation machines with highly turbulized flows: Creation of software complex for technological design. IOP Conf Ser Mater Sci Eng 776(1):012018

Artyukhov A, Krmela J, Krmelova V (2020b) Manufacturing of vortex granulators: simulation of the vortex fluidized bed functioning under the disperse phase interaction in the constrained motion. Manuf Technol 20(5):547–553

Artyukhova N (2020) Morphological features of the nanoporous structure in the ammonium nitrate granules at the final drying stage in multistage devices. J Nano- Electron Phys 12(4):04036–1–04036-6

Artyukhova N, Krmela J (2019) Nanoporous structure of the ammonium nitrate granules at the final drying: the effect of the dryer operation mode. J Nano- Electron Phys 11(4):04006–1–04006-4

Bilan Y, Pimonenko T, Starchenko L (2020) Sustainable business models for innovation and success: bibliometric analysis. E3S web of conferences, 159

Chen Q, Zhai Z, Wang L (2007) Computer modeling of multiscale fluid flow and heat and mass transfer in engineered spaces. Chem Eng Sci 62(13):3580–3588

Gibilaro LG (2001) Fluidization-dynamics. The formulation and applications of a predictive theory for the fluidized state. Butterworth-Heinemann, Woburn

Gidaspow D (1994) Multiphase flow and fluidization: continuum and kinetic theory descriptions with applications. Academic Press, San Diego

Horio M (2013) Overview of fluidization science and fluidized bed technologies. Woodhead publishing series in energy, fluidized bed Technologies for Near-Zero Emission Combustion and Gasification, 3–41

Kasych A, Vochozka M (2017) Theoretical and methodical principles of managing enterprise sustainable development. Mark Manag Innov 2:298–305

Kobushko I, Jula O, Kolesnyk M (2017) Improvement of the mechanism of innovative development of small and medium-sized enterprises. SocioEcon Chall 1(1):60–67

Koshkarev S, Evtushenko A, Roschin P (2017) Modeling of cleaning of dust emission' in fluidized bed building aspiration' collector. MATEC Web Conf 106:07020

Kosowska-Golachowska M, Luckos A, Kijo-Kleczkowska A et al (2019) Analysis of pollutant emissions during circulating fluidized bed combustion of sewage sludge. J Phys Conf Ser 1398: 012010

Krmela J, Artyukhova N, Artyukhov A (2020) Investigation of the convection drying process in a multistage apparatus with a differential thermal regime. Manuf Technol 20(4):468–473

Kuo JH, Lin CL, Ho CY et al (2021) Influence of different fluidization and gasification parameters on syngas composition and heavy metal retention in a two-stage fluidized bed gasification process. Environ Sci Pollut Res 28:22927

Kwauk M (1992) Fluidization: idealized and bubbleless, with application. Science Press, Beijing

Lipkova L, Braga D (2016) Measuring commercialization success of innovations in the EU. Mark Manag Innov 4:15–30

Litster J, Ennis B (2004) The science and engineering of granulation processes. Springer

Liu K, Gu XG, Ba DC et al (2012) Numerical research on flow characteristics of vortex stage in dry high vacuum pump. Phys Procedia 32:127–134

Mazza G, Soria J, Gauthier D et al (2016) Environmental friendly fluidized bed combustion of solid fuels: a review about local scale modeling of char heterogeneous combustion. Waste Biomass Valoriz 7:237–266

Melnyk OS (2016) Alternative design for electrocoagulation treatment of cromium-containing electroplating wastewater. J Water Chem Technol 38(1):45–50

Melnyk O, Kovalenko V, Kotok V (2020) Combination of electrocoagulation and flotation technologies in apparatus for treatment of electroplating wastewater. ARPN J Eng Appl Sci 15(22): 2639–2646

Münch J, Trieflinger S, Lang D (2019) Product roadmap – from vision to reality: a systematic literature review. In: 2019 IEEE international conference on engineering, technology and innovation, 1–8

Muralidhar P, Bhargav E, Sowmya C (2016) Novel techniques of granulation: a review. Int Res J Pharm 7(10):8–13

New York State Energy Research and Development Authority (NYSERDA) (2020) TRL calculator. Retrieved from https://portal.nyserda.ny.gov/servlet/servlet.FileDownload?file=00 Pt000000ASeCMEA1

Obodiak V, Artyukhova N, Artyukhov A (2020) Calculation of the residence time of dispersed phase in sectioned devices: theoretical basics and software implementation. Lecture notes in mechanical engineering, pp 813–820

Parikh D (2009) Handbook of pharmaceutical granulation technology, 3rd edn. Informa Healthcare

Patel P, Telange D, Sharma N (2011) Comparison of different granulation techniques for lactose monohydrate. Int J Pharm Sci Drug Res 3:222–225

Phaal R, Farrukh C, Probert D (2004) Technology roadmapping – a planning framework for evolution and revolution. Technol Forecast Soc Chang 71:5–26

Prokopov MG, Levchenko DA, Artyukhov AE (2014) Investigation of liquid-steam stream compressor. Appl Mech Mater 630:109–116

Saikh MA (2013) A technical note on granulation technology: a way to optimise granules. Int J Pharm Sci Rev Res 4:55–67

Solanki H, Basuri T, Thakkar J et al (2010) Recent advances in granulation technology. Int J Pharm Sci Rev Res 5(3):48–54

Spremberg E, Tykhenko V, Lopa L (2017) Public-private partnership in the implementation of national environmental projects. SocioEcon Chall 1(4):73–81

Stahl H (2004) Comparing different granulation techniques. Pharm Technol Eur 11:23–33

Stahl H (2010) Comparing granulation methods/Hürth: GEA pharma systems. GEA Pharma Systems, Hürth

Tahvildari K, Tavakoli H, Mashayekhi A et al (2016) Modeling and simulation of membrane separation process using computational fluid dynamics. Arab J Chem 9(1):72–78

Trojan M (2015) Computer modeling of a convective steam superheater. Arch Thermodyn 36(1): 125–137

Wang Q, Feng Y, Lu J et al (2015) Numerical study of particle segregation in a coal beneficiation fluidized bed by a TFM–DEM hybrid model: influence of coal particle size and density. Chem Eng J 260:240–257

Yang WC (2003) Handbook of fluidizaition and fluid-particle systems. Marcel Dekker, New York

Zhong W, Jin B, Zhang Y et al (2008) Fluidization of biomass particles in a gas–solid fluidized bed. Energy Fuel 22:4170–4176

Chapter 7
Vortex Granulators in Chemical Engineering: Environmental Aspects and Marketing Strategy of Implementation

Artem Artyukhov, Nadiia Artyukhova, Jan Krmela, Tetiana Vasylieva, Serhiy Lyeonov, and Olena Melnyk

7.1 Introduction

The chemical industry has always been under the close attention of organizations monitoring the environmental safety of production facilities. Due to the large volumes of products, the chemical industry faces a large volume of waste that needs to be disposed of. If we talk about the definition of "disposal," it is interpreted as "burial" (moreover, both in the bowels of the Earth with the transition to water and the atmosphere) and as a way of reuse. The second way is more promising and consistent with the sustainable development goals, despite the increase in capital costs for the new production facilities or the modernization of existing ones, as it marks in "Transforming Our World: The 2030 Agenda for Sustainable Development" (United Nations 2015). The proper sustainable development goals describe the relevance of this problem in detail regarding SDG reports (United Nations

---

A. Artyukhov (✉)
University of Economics in Bratislava, Bratislava, Slovakia

Sumy State University, Sumy, Ukraine
e-mail: a.artyukhov@pohnp.sumdu.edu.ua

N. Artyukhova · T. Vasylieva · S. Lyeonov
Sumy State University, Sumy, Ukraine
e-mail: n.artyukhova@pohnp.sumdu.edu.ua; tavasilyeva@biem.sumdu.edu.ua; s.lieonov@uabs.sumdu.edu.ua

J. Krmela
Alexander Dubcek University of Trencin, Puchov, Slovakia
e-mail: jan.krmela@fpt.tnuni.sk

O. Melnyk
Sumy National Agrarian University, Sumy, Ukraine
e-mail: olena.melnyk@snau.edu.ua

© The Author(s), under exclusive license to Springer Nature Switzerland AG 2023
S. Boichenko et al. (eds.), *Sustainable Transport and Environmental Safety in Aviation*, Sustainable Aviation, https://doi.org/10.1007/978-3-031-34350-6_7

2019a), Sustainable Development Goals report (United Nations 2019b), financing for sustainable development report (United Nations 2019c), General Global Sustainable Development Report (United Nations 2019d), Eurostat data (Eurostat 2021) and Financing for Sustainable Development Report (United Nations 2020).

Chemical production waste is highly harmful to air and water basins. At the same time, they can be successfully reused in technological conditions and design solutions for trapping. Granulation as a technique of forming a wide range of materials from solutions and melts has become widespread in the chemical engineering applications (Litster and Ennis 2004), theoretical basics (Salman et al. 2006), pharmaceutical engineering applications (Solanki et al. 2010), theoretical basics (Saikh 2013), practical implementation (Muralidhar et al. 2016), food engineering applications (Pathare and Byrne 2011), and theoretical basics (Li et al. 2017) industries. Several studies on the environmental safety of chemical production in general water pollution (Posthuma et al. 2020), green chemistry (Hao et al. 2017; Wernet et al. 2011), electrocoagulation (Melnyk 2016), filtration (Melnyk et al. 2020) and granulation technology (Mieldažys et al. 2019) show an urgent need to improve individual production stages and ensure their relationship (recycling and reuse of flows) (Khan et al. 2016).

Production of the granular products, particularly ammonium nitrate (for the agricultural sector and as an industrial explosive), is followed by significant emissions of the small fraction (dust) of ammonium nitrate, ammonia, and nitrogen oxides into the atmosphere. Recycling this waste reduces the number of raw materials because about 70% of production costs are the expenditures for energy and raw materials in the chemical industry.

Based on the theoretical description of vortex granulators' new constructions (Artyukhov et al. 2017a) and different types of fluidized bed functioning (Artyukhov and Artyukhova 2018), technological basics of granulation in vortex flows (Artyukhov and Artyukhova 2019), experimental studies of the main technological equipment of the process to obtain NH_4NO_3 with nanoporous structure (Artyukhov et al. 2017b), obtaining multilayer modified NH_4NO_3 granules with 3D nanoporous structure production (Artyukhov and Sklabinskyi 2017), practical implementation of technology of 3D nanostructured porous surface layer in NH_4NO_3 granules obtaining (Artyukhov et al. 2017c), creation of software complex for technological design (Artyukhov et al. 2020a), implementation of porous ammonium nitrate producing process in vortex granulators (Artyukhov and Sklabinskyi 2013), simulation of the vortex fluidized bed functioning (Artyukhov et al. 2020b), the authors of this work substantiate the introduction of the new granulators – a vortex granulator with a fluidized bed – into the ammonium nitrate production technology. Such granulators can significantly increase the granulation process' specific (per unit volume) productivity in the active hydrodynamic modes. A couple of works have confirmed the successful application of active hydrodynamic modes (process to obtain NH_4NO_3 with nanoporous structure (Artyukhova et al. 2020), convection drying process (Krmela et al. 2020), sectioned devices (Obodiak et al. 2020), ejectors (Prokopov et al. 2014)), describing the drying processes in the multistage device.

The main advantages of vortex granulators are as follows (Artyukhov and Ivaniia 2017):

- The compactness of the granulator due to the concurrent holding of the main stages in the production of the porous granules in the working space, for example, the humidification of the seeding agent, dehydration of the generated granules.
- The ability to monitoring the residence period of the granule in the granulator.
- Classification of granules by dimension (diameter) in the workspace of the granulator, increasing in the monodispersity degree of the commodity fraction.
- Uniformity of interaction between the high-potential heat transfer agent and granules in the vortex fluidized bed mode.
- No overheating of the granules, and reducing thermal stresses in the granules, the risk of bursting and breakdown of the core due to a clear separate introduction of heat transfer agents with different enthalpies in separate areas of the spiral upper part of the device.
- Uniform interaction of the wet granules with the heat transfer agent's stream in the vortex fluidized bed mode, the minimum required time to form a honeycombed superficial layer on the granules.
- Improving the quality of the finished product and its monodispersity degree due to the differentiated interaction of the granules with the heat transfer agent's flow at each granulation stage.
- The opportunity to obtain granules of vast fractions with a high monodispersity extent of each fraction.

However, one should note that these advantages in the vortex granulators directly relate to the target process. It is necessary to improve their design to increase the environmental safety of the vortex granulators' operation.

An essential stage in introducing new technology and design solutions is creating a marketing strategy to enter improved devices and technologies into the market. According to several studies (environmental management and green brand (Starchenko et al. 2021), sustainable business models for innovation (Bilan et al. 2020), measuring commercialization success of innovations (Lipkova and Braga 2016), strategic planning (Andrade and Loureiro 2020), partnership in the implementation of national environmental projects (Spremberg et al. 2017)), at this stage, it is necessary to develop roadmaps for the introduction of technologies and technological trajectories to choose the optimal solution (structure) for the environmental safety of production as well as energy efficiency actions (managing sustainable enterprise development (Kasych and Vochozka 2017), energy-efficient innovations (Panchenko et al. 2020), and eco-innovations (Lesakova 2019)). The article aims to develop new methods for the disposal of waste gases in granulating plants using vortex granulators and to create a technological trajectory (an algorithm for promoting equipment as a product) for the introduction of new designs of vortex granulators.

7.2 Constructive Solutions to Improve the Environmental Safety of the Production of Granular Products in Vortex Granulators

Authors propose the following essential areas of disposal of harmful emissions of ammonium nitrate production and devices to improve the environmental safety of granulation plants:

1. Trapping fine fraction and dirt with subsequent sending for preparation of the solution – separation stages (Artyukhov et al. 2020b).
2. Trapping the ammonia with the formation of ammonia water for industry purposes – vortex contact heat and mass transfer stages (Artyukhov 2014).
3. Trapping the small fraction with the subsequent sending to further growing – takes place directly in the vortex granulator (Artyukhov and Sklabinskyi 2013).

This section represents the following options for outgoing gas cleaning after the granulation stage:

1. Classification of granules by size with subsequent selection of small fraction.
2. Installation of mass transfer-separation vortex contact stages.

Classification of Granules by Size with Subsequent Selection of the Small Fraction Granulation devices with a permanent cross-sectional space do not fully ensure granule classification and separation of the non-commodity granules in the device. It means that in the vortex granulator working volume, the constant ascending velocity of the gas flow is maintained, which corresponds to the working velocity of the dispersed phase (or the fraction of dispersed phase in a wide range). It is possible to classify granules with a permanent cross-sectional space introducing gas into the device in several streams with the position of the injection sites at various altitudes. This way of classification is very energy-consuming and has not been widely used yet.

A better way of classifying the dispersed material is to implement granulators with a variable cross-sectional space of the workspace. Due to the creation of the velocity element fields in the gas flow along with the device's height, various hydrodynamic modes for the granule motion are built. Granules of various diameters (under the conditions of classified granules of the same material) or various masses (in terms of the form of porous granules or multilayer granules) are distributed by the device's height. It allows to obtain a product of a given quality and to separate fine particles. Fine particles can be used as a seeding agent.

The classification is realized on the background of the mathematical model of granules (Artyukhov and Sklabinskyi 2015a) and gas flow (Artyukhov and Sklabinskyi 2015b) motion and the experimental research results using the original software classification in vortex flow©. The basic stages of the calculation are given in Figs. 7.1, 7.2, 7.3, 7.4 and 7.5.

7 Vortex Granulators in Chemical Engineering: Environmental Aspects...

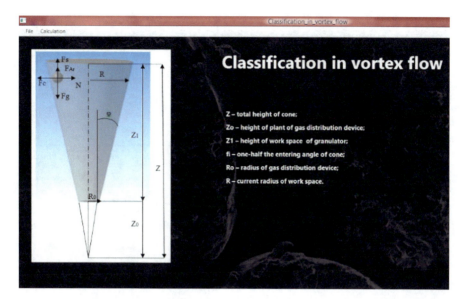

Fig. 7.1 The working volume of the device

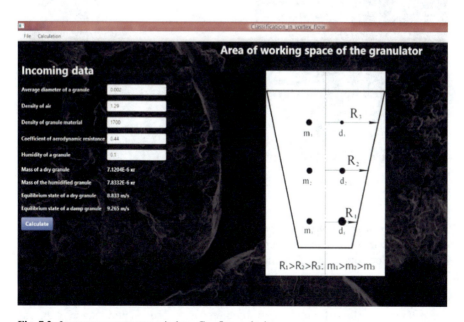

Fig. 7.2 Input parameter entry window. Gas flow velocity

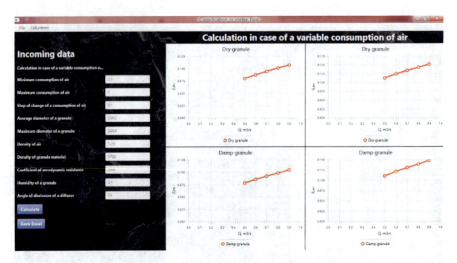

Fig. 7.3 Calculation of the geometry of the granulator's working volume at a variable airflow rate

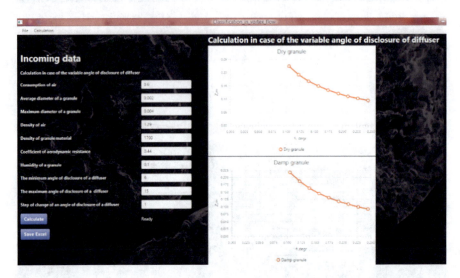

Fig. 7.4 Calculation of the vortex granulator's geometry at a variable angle of diffuser disclosing

Installation of Mass Transfer-Separation Vortex Contact Stages It is proposed to use a vortex mass transfer contact section and a vortex plate with mass transfer-separation elements to trap dust and ammonia from the outgoing gases.

Equipping the vortex granulator with an inertial-filter vortex separation section (Fig. 7.6) enables inertially separating small-fraction granules and creating a vortex foamed water-air layer for hydro dust filtration and absorption purification of the gas flow. The vortex gas flow energy from the workspace utilizes the exhaust heat

7 Vortex Granulators in Chemical Engineering: Environmental Aspects... 113

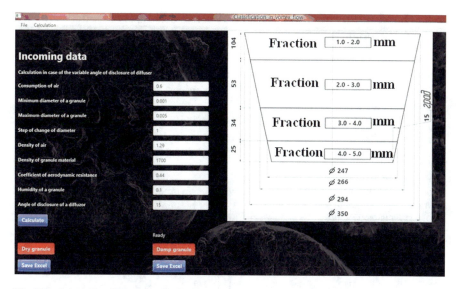

Fig. 7.5 Calculation of the granule distribution by fractions in the granulator

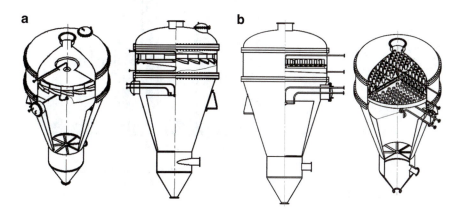

Fig. 7.6 Vortex granulators: (**a**) with vortex mass transfer contact section; (**b**) with a vortex tray with mass transfer and separation elements

transfer agent's energy. It captures highly dispersive dust particles by hydro filtration and absorptive purification of the exhaust heat transfer agent's gas flow from harmful gas.

The device outfitting with a vortex mass transfer contact section with a vortex plate with mass transfer and separation elements allows to carry out absorption purification of outgoing gases, directly in the vortex granulator, when exhaust heat transfer agent passes mass transfer elements in which liquid captures ammonia vapor.

7.3 Evaluation of the New Granulator's Construction Efficiency

Because of the classification and further trapping of small granules and dirt in the mass transfer sections, the fractional compound of the granules is significantly increased (Fig. 7.7).

The results of experimental studies in Fig. 7.8 and Fig. 7.9 show the vortex granulator efficiency in the classification and vortex stages of outgoing gases purification from dirt and ammonia. The comparative analysis regarding the

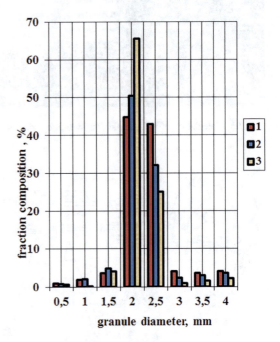

Fig. 7.7 The fractional compound of particles in a vortex granulator: 1, vortex granulator; 2, vortex granulator with vortex mass transfer contact section; 3, vortex granulator with vortex plate with mass transfer and separation elements

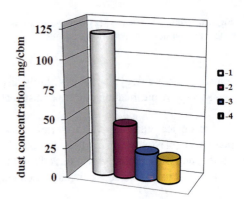

Fig. 7.8 The middle dirt content of ammonium nitrate in the gases leaving the granulator: 1, the fluidized bed granulator; 2, vortex granulator; 3, vortex granulator with vortex plate; 4, vortex granulator with vortex mass transfer section

7 Vortex Granulators in Chemical Engineering: Environmental Aspects...

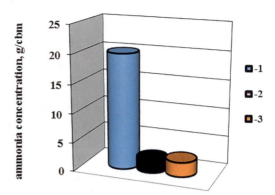

Fig. 7.9 The average content of ammonia in the gases leaving the granulator: 1, vortex granulator; 2, vortex granulator with vortex plate; 3, vortex granulator with vortex mass transfer section

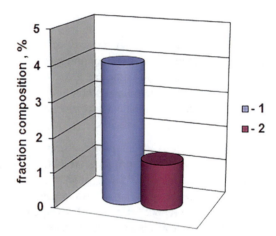

Fig. 7.10 Mass fraction of granules larger than 3.0 mm in the finished product: 1, granulator with a classic weighted layer; 2, vortex granulator

granulometric composition of the product obtained using fluidized bed granulators and small vortex devices of different designs (Fig. 7.10, 7.11) shows that a product obtained using vortex granulators has a greater degree of monodispersity. Analysis of the gas composition when leaving the vortex granulator shows the advantage of this equipment over granulators with a classical fluidized bed. Additional installation of separation devices significantly reduces the dust content in the outgoing gases.

7.4 New Technology Map and Trajectory of Their Implementation

We single out granulation in devices with a fluidized bed as a key process in the basic technology – chemical engineering, the industry of obtaining a granular product. This technology has become widespread in the chemical industry but

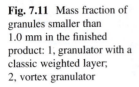
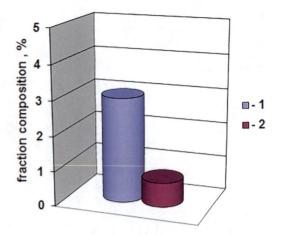

Fig. 7.11 Mass fraction of granules smaller than 1.0 mm in the finished product: 1, granulator with a classic weighted layer; 2, vortex granulator

requires modernization to increase specific productivity. This technology is not optimal for getting porous ammonium nitrate granules because it does not allow to control the humidification time and heat treatment of granules. As a trial technology, a method for producing porous ammonium nitrate in a vortex fluidized bed is used. Vortex granulators do not have the disadvantages listed above and can reduce energy costs for the granulation process.

Technology Map Analysis of the current market for porous ammonium nitrate producers shows that the products are obtained in granulation towers. This type of equipment is characterized by significant capital costs for producing, maintenance, and repair since the granulation towers have a large diameter (up to 16 m) and are sufficiently high (about 30–50 m). The large dimensions of tower-type granulation equipment also determine production and operation complexity. Besides, the granulation towers have a comparatively low specific productivity.

The fluidized bed granulation technology with a pilot granulation in vortex devices is proposed as a key technology. It will provide the bulk industrial explosives market for blasting operations in open-pit and underground mines. Twenty patents protect the technology and vortex granulators' industrial designs, ten copyright certificates protect new software products for the granulation process calculation. The development of granulation technology in vortex granulators is shown in Fig. 7.12.

Technological Trajectory The Sumy State University research laboratory of granulation and heat and mass transfer equipment carried out comparative studies of various methods to obtain granular products: nodulizing, pressing crystallization during the cooling, and fluidization.

Analysis of the modern market of granular product producers for different industries shows that fluidized bed granulation is the latest (advanced) granulation among the listed methods. Thus, the ammonium nitrate granulation technology is

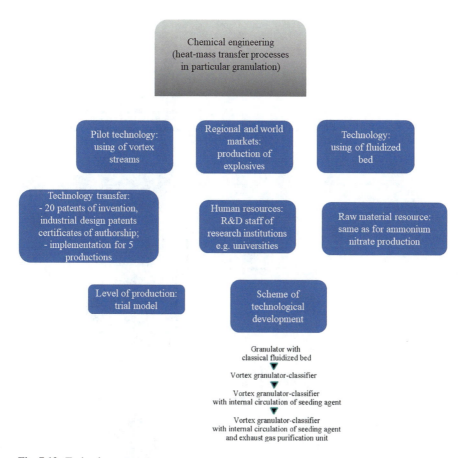

Fig. 7.12 Technology map

further developed using fluidized bed granulation technology. Four types of fluidized bed granulation equipment are considered: classical fluidized bed granulators, rotor granulators (rotating disc granulators), granulators-classifiers, and vortex granulators.

According to the comparative analysis of the benefits and lacks of the equipment for further developing an industrial sample of the granulator, two types of equipment are selected: granulators-classifiers and vortex granulators. Three types of devices are investigated: a vortex granulator-classifier, a vortex granulator-classifier with the internal circulation of seeding agent, a vortex granulator-classifier with the internal circulation of seeding agent and the cleaning unit for outgoing gases. The third type is selected for the market as the most energy-efficient and environmentally friendly (Fig. 7.13).

Fig. 7.13 Technological trajectory

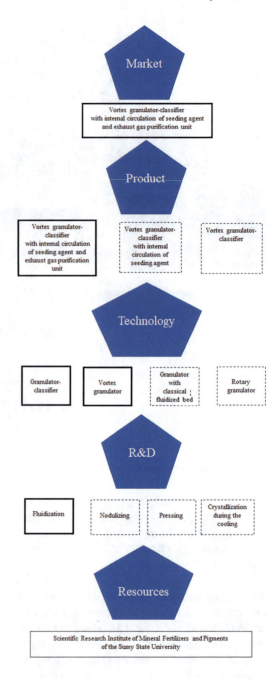

7.5 Conclusions

The research results regarding the outgoing gas composition and the fractional composition of the granular product enable us to assert the effectiveness of the proposed solutions. The investigated map of the granulation technology and the technological trajectory to choose the optimal granulation method ensures the effective execution of new granulation technologies in vortex devices with the harmful emission disposal beginning from making constructive solutions to their implementation. Description of the technology transfer stage is an obligatory part of the new technology development or improving current ones.

Acknowledgments The authors thank researchers at the Marketing Department, Sumy State University, Department of Numerical Methods and Computational Modeling, and Department of Materials Technologies and Environment, Alexander Dubcek University of Trenčín, for their valuable comments during the article preparation.

This research work had been supported by the Cultural and Educational Grant Agency of the Slovak Republic (KEGA) under the project "Simulations of Basic and Specific Experiments of Polymers and Composites Based on Experimental Data in Order to Create a Virtual Computational-Experimental Laboratory for Mechanical Testing" (No. KEGA 003TnUAD-4/2022), by the Scientific Grant Agency of the Ministry of Education, Science, Research, and Sport of the Slovak Republic, and by the Slovak Academy of Sciences under the project "Changes in the Approach to the Creation of Companies' Distribution Management Concepts Influenced by the Effects of Social and Economic Crises Caused by the Global Pandemic and Increased Security Risks" (VEGA 1/0392/23) and under the project "Advancement and Support of R&D for 'Centre for Diagnostics and Quality Testing of Materials' in the Domains of the RIS3 SK specialization," code NFP313011W442.

References

Andrade HS, Loureiro GA (2020) Comparative analysis of strategic planning based on a systems engineering approach. Bus Ethics Leadersh 4(2):86–95

Artyukhov A (2014) Optimization of mass transfer separation elements of columnar equipment for natural gas preparation. Chem Pet Eng 49(11–12):736–741

Artyukhov A, Artyukhova N (2018) Utilization of dust and ammonia from exhaust gases: new solutions for dryers with different types of fluidized bed. J Environ Health Sci Eng 16(2): 193–204

Artyukhov A, Artyukhova N (2019) Technology and the main technological equipment of the process to obtain NH_4NO_3 with nanoporous structure. Springer Proc Phys 221:585–594

Artyukhov AE, Ivaniia AV (2017) Obtaining porous ammonium nitrate in multistage and multifunctional vortex granulators. Naukovyi Visnyk Natsionalnoho Hirnychoho Universytetu 6:68–75

Artyukhov A, Sklabinskyi V (2013) Experimental and industrial implementation of porous ammonium nitrate producing process in vortex granulators. Naukovyi Visnyk Natsionalnoho Hirnychoho Universytetu 6:42–48

Artyukhov A, Sklabinskyi V (2015a) Theoretical analysis of granules movement hydrodynamics in the vortex granulators of ammonium nitrate and carbamide production. Chem Chem Technol 9(2):175–180

Artyukhov A, Sklabinskyi V (2015b) Hydrodynamics of gas flow in small-sized vortex granulators in the production of nitrogen fertilizers. Chem Chem Technol 9(3):337–342

Artyukhov A, Sklabinskyi V (2017) Investigation of the temperature field of coolant in the installations for obtaining 3D nanostructured porous surface layer on the granules of ammonium nitrate. J Nano- Electron Phys 9(1):01015-1–01015-4

Artyukhov A, Artyukhova N, Ivaniia A et al (2017a) Progressive equipment for generation of the porous ammonium nitrate with 3D nanostructure. In: Proceedings of the 2017 IEEE 7th international conference on nanomaterials: applications and properties

Artyukhov A, Artyukhova N, Ivaniia A et al (2017b) Multilayer modified NH_4NO_3 granules with 3D nanoporous structure: effect of the heat treatment regime on the structure of macro-and mesopores. In: Proceeding of the IEEE international young scientists forum on applied physics and engineering, pp 315–318

Artyukhov A, Obodiak V, Boiko P et al (2017c) Computer modeling of hydrodynamic and heat-mass transfer processes in the vortex type granulation devices. CEUR Workshop Proc 1844:33–47

Artyukhov A, Artyukhova N, Krmela J et al (2020a) Granulation machines with highly turbulized flows: creation of software complex for technological design. IOP Conf Ser Mater Sci Eng 776(1):012018

Artyukhov A, Krmela J, Krmelova V (2020b) Manufacturing of vortex granulators: simulation of the vortex fluidized bed functioning under the disperse phase interaction in the constrained motion. Manuf Technol 20(5):547–553

Artyukhova N, Krmela J, Krmelova V (2020) Quality indicators of ammonium nitrate with Nanoporous surface structure: final drying stage. In: Proceedings of the 2020 IEEE 10th international conference on "nanomaterials: applications and properties," 9309583

Bilan Y, Pimonenko T, Starchenko L (2020) Sustainable business models for innovation and success: bibliometric analysis. E3S Web of conferences,159: 04037

Eurostat (2021) Sustainable development in the European Union. https://eceuropaeu/eurostat/en/web/products-statistical-books/-/ks-03-21-096. Accessed 29 Jan 2022

Hao Q, Jinping T, Xing L et al (2017) Using a hybrid of green chemistry and industrial ecology to make chemical production greener. Resour Conserv Recycl 122:106–113

Kasych A, Vochozka M (2017) Theoretical and methodical principles of managing enterprise sustainable development. Mark Manag Innov 2:298–305

Khan AR, Al-Awadi L, Al-Rashidi MS (2016) Control of ammonia and urea emissions from urea manufacturing facilities of petrochemical industries company (PIC). J Air Waste Manag Assoc 66(6):609–618

Krmela J, Artyukhova N, Artyukhov A (2020) Investigation of the convection drying process in a multistage apparatus with a differential thermal regime. Manuf Technol 20(4):468–473

Lesakova L (2019) Small and medium enterprises and eco-innovations: empirical study of Slovak SMEs. Mark Manag Innov 3:89–97

Li Z, Luo J, Jiang Q et al (2017) Roles of the Main physical properties of the wet granulation product of hawthorn leaf extract mixtures in high shear granulation. J Food Process Preserv 41

Lipkova L, Braga D (2016) Measuring commercialization success of innovations in the EU. Mark Manag Innov 4:15–30

Litster J, Ennis B (2004) The science and engineering of granulation processes. Springer, Basel

Melnyk O (2016) Alternative design for electrocoagulation treatment of chromium-containing electroplating wastewater. J Water Chem Technol 38(1):45–50

Melnyk O, Kovalenko V, Kotok V et al (2020) Combination of electrocoagulation and flotation technologies in apparatus for treatment of electroplating wastewater. ARPN J Eng Appl Sci 15(22):2639–2646

Mieldažys R, Jotautienė E, Jasinskas A et al (2019) Investigation of physical-mechanical properties and impact on soil of granulated manure compost fertilizers. J Environ Eng Landsc 27(3): 153–162

Muralidhar P, Bhargav E, Sowmya C (2016) Novel techniques of granulation: a review. Int Res J Pharm 7(10):8–13

Obodiak V, Artyukhova N, Artyukhov A (2020) Calculation of the residence time of dispersed phase in sectioned devices: theoretical basics and software implementation. Lecture Notes in Mech Eng. pp 813–820

Panchenko V, Yu H, Ya U et al (2020) Energy-efficient innovations: marketing, management and law supporting. Mark Manag Innov 1:256–264

Pathare P, Byrne E (2011) Application of wet granulation processes for granola breakfast cereal production. Food Eng Rev 3:189–201

Posthuma L, Zijp M, De Zwart D et al (2020) Chemical pollution imposes limitations to the ecological status of European surface waters. Sci Rep 10:14825

Prokopov MG, Levchenko DA, Artyukhov AE (2014) Investigation of liquid-steam stream compressor. Appl Mech Mater 630:109–116

Saikh M (2013) A technical note on granulation technology: a way to optimize granules. Int J Pharm Sci Rev Res 4:55–67

Salman A, Hounslow M, Seville JP (2006) Granulation. Elsevier Science Ltd, Amsterdam

Solanki H, Basuri T, Thakkar J et al (2010) Recent advances in granulation technology. Int J Pharm Sci Rev Res 5(3):48–54

Spremberg E, Tykhenko V, Lopa L (2017) Public-private partnership in the implementation of national environmental projects. SocioEcon Chall 1(4):73–81

Starchenko L, Lyeonov S, Vasylieva T et al (2021) Environmental management and green brand for sustainable entrepreneurship. E3S Web of Conferences, 234

United Nations (2015) Transforming our world: the 2030 agenda for sustainable development. https://sdgsunorg/2030agenda. Accessed 29 Jan 2022

United Nations (2019a) Leaving biodiversity, peace and social inclusion behind SDG preferences in the World's five major SDG reports. Basel Institute of Commons and Economics, Basel

United Nations (2019b) The sustainable development goals report 2019. https://unstatsunorg/sdgs/report/2019/The-Sustainable-Development-Goals-Report-2019pdf. Accessed 29 Jan 2022

United Nations (2019c) Report of the inter-agency task force on financing for development 2019: financing for sustainable development report 2019. https://developmentfinanceunorg/fsdr2019. Accessed 29 Jan 2022

United Nations (2019d) General, global sustainable development report 2019: the future is now – science for achieving sustainable development. https://sustainabledevelopmentunorg/globalsdreport/2019. Accessed 29 Jan 2022

United Nations (2020) Financing for sustainable development report 2020. https://ec.europa.eu/eurostat/en/web/products-statistical-books/-/ks-03-21-096. Accessed 29 Jan 2022

Wernet G, Mutel C, Hellweg S et al (2011) The environmental importance of energy use in chemical production. J Ind Ecol 15:96–107

Chapter 8
Evaluation of Automobile Road Construction Environmental and Economic Efficiency Based on Public-Private Partnership

Yevheniia Tsiuman, Mykola Tsiuman, and Anatolii Morozov

Nomenclature

SB   State budget
T   Taxes
ERI   Enterprises of related industries
M   Materials
E_q Equipment
CRTS   Consumers of road transport services
I_n Income
GS   Gas stations
MS   Microsoft

8.1 Introduction

Automobile roads are the primary transport infrastructure branch that provides goods and passengers transportation by automobile vehicles. Unsatisfactory road condition reduces transportation speed and safety, which harms other economic sector efficiency. Therefore, road development and bringing their quality to international standards are the most urgent task. The solution of this task will help improve the economic situation in the country.

Automobile roads are state property. Therefore, the state budget is the primary funding source for their construction and maintenance. However, the state budget

---

Y. Tsiuman (✉) · M. Tsiuman · A. Morozov
National Transport University, Kyiv, Ukraine
e-mail: y.tsiuman@ntu.edu.ua

© The Author(s), under exclusive license to Springer Nature Switzerland AG 2023
S. Boichenko et al. (eds.), *Sustainable Transport and Environmental Safety in Aviation*, Sustainable Aviation, https://doi.org/10.1007/978-3-031-34350-6_8

limitation does not allow for total financing of road construction projects. This fact encourages to search for private investors. A public-private partnership is the most common form of investment activity in the road construction field in the world. At the same time, an essential condition for the successful public-private partnership implementation in the services field is the economic feasibility for partners and potential services consumers.

Numerous domestic and foreign scientists had dedicated their works to issues of application of the different mechanisms for cooperation between the state and private business, institutional aspect research of these relations, and their impact on the economic development. They are A.V. Bazyliuk, V.H. Varnavskyi, I.V. Zapatrina, Yu.S. Vdovenko, O.V. Zhulyn, and others (Bazyliuk and Zhulyn 2007; Varnavskyi 2011; Zapatrina 2010; Vdovenko 2008; Zhulyn 2009).

At the same time, automobile vehicles are a source of roadside pollution during road transportation. This pollution level depends on traffic flow composition moving on the road. It also depends on the traffic flow speed and intensity determined by the road technical parameters. Numerous works of Gutarevych Y.F., Mateychyk V.P., Kanylo P.M., and other scientists (Gutarevych et al. 2006; Mateichyk 2006; Kanylo 2013) are devoted to research harmful influence of the transport on the environment. In these works, methods of decreasing this harmful influence are also proposed.

However, the issues of road construction efficiency comprehensive environmental and economic analysis, taking into account the impact of the public-private relations, are still insufficiently studied. It determines the relevance of this study.

Therefore, the purpose of this study is to evaluate environmental and economic efficiency of road construction based on public-private partnership.

8.2 Public-Private Partnership Terms

Automobile road development is closely linked to investment activity. The finance sources of investment activity include the state, private, and foreign ones. They are the state road fund formed by excise and customs payments from the circulation of fuels and vehicles, and loans from international financial organizations, foreign direct investment and other. However, the foreign investment attracting activity is primarily influenced by political factors and institutional imperfections in government regulation. Therefore, private sources cause the most significant interest for investment activity financing. The private investment sources are also the basis of public-private partnership, which is seen as one of the tools to intensify investment activity.

Public-private partnership provides a contractual commercial operation of the state object by the private partner. In essence, a public-private partnership is a system of relationships between the state, private partner, and consumers, each of which has its own goal. These goals are achieved creating regulatory and legal preconditions for attraction of private investment, protecting consumer interests. Also it is sharing risk, reducing the state budget burden, using quality services and services payment, and increasing trust in public institutions. Thus, the successful implementation of a public-private partnership depends on mutually beneficial conditions created for its participants.

The main legal, financial, and organizational principles of public-private partnership subject interaction in the road construction field are defined at the legislative level and have a sufficiently high implementation level of international trends. Also, the automobile road development using investment resources is one of the priorities in the "Transport Strategy of Ukraine Until 2030" government program for investment activities development.

A concession is the most appropriate form of public-private partnership in the road construction field. It is related with the protection of partner interest, investment distribution between them, mechanism of investment return, and protection against the profit loss risk. A concession provides new object creation or significant modernization of existing objects. The lot of proposed mechanisms of public-private partnership in the road construction field does not take into account the consumer interest. In particular, these mechanisms do not guarantee income to the concessionaire due to reduced demand on toll road. In addition, these mechanisms do not consider the environmental effects of toll road construction projects. These facts necessitate the creation of a more effective mechanism for intensifying investment activity in the road construction field on a public-private partnership basis.

The analysis shows that available funding sources can be used only to maintain the existing roads. However, it is practically impossible to use for necessary road construction project implementation. The growth of finance sources for investment activity is limited by the investment attractiveness of road construction objects. It also depends on institutional, economic, social, and psychological factors.

8.3 Research Methodology

In order to analyze the environmental and economic efficiency of road construction on a public-private partnership basis, a model based on a system approach is proposed (Fig. 8.1).

The model has a four-level structure and describes the system interaction between the road construction and maintenance processes by road construction enterprises. The model also describes the road operation by road transport enterprises and individual car owners. Finance resources and transportation volumes are the input of process. The output process parameters are transport and operational performance of the road. Also, they are social, economic, and environmental effects of the road operation. Feedbacks maximize the output parameters by controlling the road construction and maintenance processes. Public and private partners and consumers form the environment related with processes of the road construction, maintenance, and operation. These relations determine the influence of the primary input, output, and process parameters on the interest of partners and consumers.

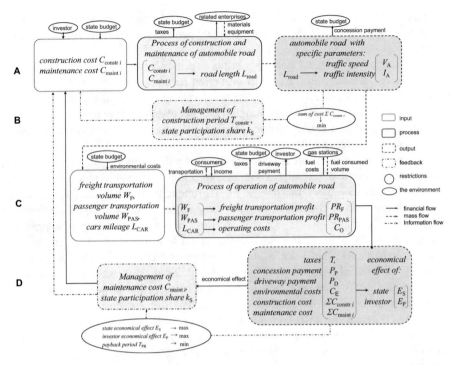

Fig. 8.1 Evaluation model of automobile road construction environmental and economic efficiency based on public-private partnership. (Adapted from Tsiuman 2020)

At the A level, the processes of road construction and maintenance are presented. This process input is the financial resources to ensure the road construction $C_{constr\ i}$ or road maintenance $C_{maint\ i}$. These resources consist of investment or state budget funds from the investor (I) or the state budget (SB). They are presented as the environment that is related to the system by finances. These financial relations are the paid taxes (T). The environment is also presented by enterprises of related industries (ERI). The ERI provide road construction enterprises with relevant materials (M) and equipment (Eq) for road construction.

The output of the A level is an automobile road of a specific category with the corresponding transport and operational performance. These performances are length L_{road}, traffic intensity I_A, and traffic speed V_A. The concession payment P_P is paid to the state budget in case of the commercial road operation by the investors on concession terms.

The system restriction is the value of the road construction costs or operation costs $\Sigma C_{constr\ i}$. It is determined by the provided transport and operational performance of the road. Implementation of this restriction is provided by B level feedback. It regulates the cost level by controlling the road construction duration T_{constr} and the state partner participation share k_S in the construction costs.

8 Evaluation of Automobile Road Construction Environmental... 127

The input of the C level is the volume of freight W_F or passenger W_{PAS} transportation and the cars mileage L_{CAR} provided by this road. These performances determine the operation process of road transport enterprises in the region. They provide services to consumers of road transport services (CRTS) and receive the corresponding income (In). Also, they are consumers of fuel volume V_{FUEL} by cost C_{FUEL} from gas stations (GS) and pay driveway payment P_D for a toll road that covers the investor's or the state budget costs. Also, they pay taxes coming into the state budget (C level process). This process performances are the enterprise profits from freight PR_F or passenger PR_{PAS} transportation and operating costs C_O of car owners. At the same time, the state budget compensates for environmental costs C_E that occur due to environmental pollution by vehicles.

The economic effect E obtained from road transport operation on the built road is distributed between the state E_S and the private E_P partner depending on their participation in the project. These effects are determined by the combination of paid taxes T, concession payments P_P, driveway payment P_D, environmental costs C_E, road construction costs $\Sigma C_{\text{constr } i}$, and road maintenance costs $\Sigma C_{\text{maint } i}$.

Feedback at the D level regulates the road maintenance costs and the state participation share in these costs to ensure the road efficient operation and its condition maintaining maximum road transport operation efficiency. It ensures maximum environmental and economic effect $E_{\max}$ by all partnership participants with minimum investment payback period $T_{PB}^{\min}$.

Thus, the potential environmental and economic effect from a toll road operation will depend on its transport and operational performance.

Determination of individual components of environmental and economic effect is carried out using the methodology developed by authors. It is based on statistical data on the construction and maintenance costs of roads with specific technical characteristics, transportation volumes, and traffic flow structure.

The resource amount spent on the road construction ($C_{\text{constr } i}$, UAH billion) in the current year is as follows:

$$C_{\text{constr } i} = i_I^{T_i - 1} \cdot \frac{C_{\text{constr}}}{T_{\text{constr}}} \cdot \left(1 + \frac{2 \cdot (2 \cdot k_C - 1) \cdot (T_{\text{constr}} - 2 \cdot T_i + 1)}{T_{\text{constr}}}\right) \qquad (8.1)$$

where i_I is inflation index

T_i is the current year of the road construction
C_{constr} is the initial cost of road construction, UAH billion
T_{constr} is road construction period, years
k_C is intensity coefficient of road construction, $k_C = 0.24$.

The k_C coefficient is introduced because the object financing is not uniform. The initial cost C_{constr} and period T_{constr} of construction are determined on the design documentation basis of a specific road. The road maintenance cost $C_{\text{maint } i}$ depends on its operation conditions and current technical condition.

The built section length of the road (L_{road}, km) within the project implementation for the current year is as follows:

$$L_{\text{road}} = L_{\text{road}}^{max} \cdot \frac{T_i}{T_{\text{constr}}} \tag{8.2}$$

where L_{road}^{max} is nominal road length within the project implementation, km.

Intensity I_A and speed V_A performances of cars will depend on the road category and determine its construction cost. According to a comparative analysis of construction cost and expected intensity of different road projects, the correlation between the expected intensity and initial construction cost was established. It is used in evaluating road construction's environmental and economic efficiency.

Average daily traffic intensity (I_A, cars per day) on certain road section is as follows:

$$I_A = \frac{\sum_{i=1}^{6} \sum_{j=1}^{7} \sum_{k=1}^{11} L_{V\ ijk}}{365 \cdot L_{road}^{max}} \tag{8.3}$$

where $L_{V\ ijk}$ is annual mileage by the road of a vehicle of the i-th category (M1...M3 or N1...N3) and the j-th environmental class (Euro 0...Euro 6), using the k-th fuel type (gasoline, diesel, gas, hybrid, electric, and others), km.

The dependencies (8.2) and (8.4) are used to determine the annual mileage $L_{V\ ijk}$. The dependence takes into account the relevant vehicle share in the flow structure:

$$L_{V\ ijk} = L_{V\ ijk}^R \cdot \frac{L_{\text{road}}^{max}}{L_{\text{road}\ \Sigma}^I} \cdot \left(1 + \frac{L_{\text{road}}}{L_{\text{road}}^{max}} \cdot \left(\frac{V_A}{V_{V\ ijk}} - 1\right)\right) \tag{8.4}$$

where $L_{V\ ijk}^R$ is the average annual mileage of vehicles of the i-th category with propulsion systems corresponding to the j-th environmental class and using the k-th fuel type in the road location region, km; $L_{\text{road}\ \Sigma}^I$ is the length of the roads in the location region of the studied road, reduced to the first category of roads, km; $L_{\text{road}\ \Sigma}^I$ is determined using the dependency (8.5); and $V_{V\ ijk}$ is the average speed of vehicles of the i-th category with propulsion systems corresponding to the j-th environmental class and using the k-th fuel type for existing roads in the region, km/h.

The average annual mileage of vehicles in the road location region is determined on the statistical data processing basis. It can be described by an approximate dependence on the current calendar year. For example, for the location region of the Great Ring Road in Kyiv, the approximate dependence for annual bus mileage has a form $L_{\text{BUS}}^R = (0,06 \cdot \ln(Y - 2000) + 0,515) \cdot 10^9$, where Y is the current year.

The road length in the studied road location region is reduced to the first category of roads ($L_{\text{road}\ \Sigma}^I$, km):

$$L_{\text{road } \Sigma}^{I} = \sum_{i=1}^{n} k_{\text{red } i} \cdot L_{\text{road } i}, \tag{8.5}$$

where $k_{\text{red } i}$ is the length reduction coefficient of the i-th category road to the first category of roads, wherein it is determined by the ratio of average daily traffic intensity for the i-th category road to the traffic intensity for the road of the first category ($k_{\text{red I}} = 1, 0$; $k_{\text{red II}} = 0, 467$; $k_{\text{red III}} = 0, 1$; $k_{\text{red IV}} = 0, 038$; $k_{\text{red V}} = 0, 005$); $L_{\text{road } i}$ is the length of region roads of the i-th category, km (statistical data); and n is the number of road categories, $n = 5$.

The amounts of taxes and concession payments are determined on a current legislation basis.

Perspective freight or passenger transportation volume by specific vehicles on the designed road (W_{Vijk}, $t$km, or pas.km; it depends on the category, the environmental class, and the fuel type) is determined using the dependency (8.4) as follows:

$$W_{Vijk} = L_{V \, ijk} \cdot M_{V \, ijk}^{F} \left(N_{V \, ijk}^{\text{PAS}} \right) \tag{8.6}$$

where $M_{V \, ijk}^{F}$ is cargo transported by a vehicle of the i-th category with propulsion systems corresponding to the j-th environmental class and using the k-th fuel type, t, and $N_{V \, ijk}^{\text{PAS}}$ is a passenger number transported by a vehicle of the i-th category with propulsion systems corresponding to the j-th environmental class and using the k-th fuel type, pas.

Road transportation profit (PR_{Vijk}, UAH billion) is based on the following dependency (8.6):

$$PR_{Vijk} = W_{Vijk} \cdot \left(T_{Vijk}^{T} - C_{Vijk}^{T} \right) \cdot \left(1 - \frac{k_{TPR}}{100} \right) \cdot 10^{-9} \tag{8.7}$$

where T_{Vijk}^{T} is the tariff for freight or passenger transportation by a vehicle of the i-th category with propulsion systems corresponding to the j-th environmental class and using the k-th fuel type, UAH/tkm, UAH/pas.km; C_{Vijk}^{T} is the cost of freight or passenger transportation by a vehicle of the i-th category with propulsion systems corresponding to the j-th environmental class and using the k-th fuel type, UAH/tkm, UAH/pas.km; and k_{TPR} is corporate income tax rate, %.

Operating costs of car owners (C_{Vijk}^{O}, UAH billion) are calculated using dependencies (8.2) and (8.4):

$$C_{Vijk}^{O} = L_{Vijk} \cdot \frac{Q_{Vijk}^{\text{FUEL}}}{100} \cdot \frac{C_{k}^{\text{FUEL}}}{\delta_{\text{FUEL}}} \cdot \left(1 - \frac{L_{\text{road}}}{L_{\text{road}}^{\max}} \cdot (1 - k_{\text{FUEL}}) \right) \cdot 10^{-9} \tag{8.8}$$

where Q_{Vijk}^{FUEL} is the fuel consumption rate for a vehicle of the i-th category with propulsion systems corresponding to the j-th environmental class and using the k-th fuel type, l/100 km; C_k^{FUEL} is the price of the k-th fuel type, UAH/l; δ_{FUEL} is the share of fuel costs in total operating costs, $\delta_{FUEL} = 0,6$; and k_{FUEL} is the coefficient of traffic conditions influence on fuel consumption, which depends on traffic intensity.

Volumes of fuel consumption by road transport (V_{FUEL}, l) are based on dependencies (8.2) and (8.4):

$$V_{\text{FUEL}} = \left(1 - (1 - k_{\text{FUEL}}) \cdot \frac{L_{\text{road}}}{L_{\text{road}}^{\text{max}}}\right) \cdot \sum_{i=1}^{6} \sum_{j=1}^{7} \sum_{k=1}^{11} \frac{Q_{Vijk}^{\text{FUEL}} \cdot L_{Vijk}}{100} \qquad (8.9)$$

The amount of excise tax income from fuel sales is determined based on consumed fuel volume following the legislation.

Total annual driveway payment by vehicle owners (P_D, UAH billion) is calculated using dependencies (8.2) and (8.4):

$$P_D = \sum_{i=1}^{6} \sum_{j=1}^{7} \sum_{k=1}^{11} T_{Vijk} \cdot L_{Vijk} \cdot \frac{L_{\text{road}}}{L_{\text{road}}^{\text{max}}} \cdot 10^{-9} \qquad (8.10)$$

where T_{Vijk} is toll road tariff (8.11) for vehicles of the i-th category with propulsion systems corresponding to the j-th environmental class and using the k-th fuel type, UAH/km.

Estimated toll road tariff (T_{Vijk}, UAH/km) is based on dependencies (8.2), (8.4), and (8.8):

$$T_{Vijk} = \frac{\Delta C_{Vijk}^O \cdot 10^9}{L_{Vijk} \cdot \frac{L_{\text{road}}}{L_{\text{road}}^{\text{max}}}} \qquad (8.11)$$

where ΔC_{Vijk}^O is the amount of operating cost reduction using the toll road for vehicles of the i-th category with propulsion systems corresponding to the j-th environmental class and using the k-th fuel type, UAH billion.

The operating cost reduction is due to the lower costs of fuel and lubricants, tires, and maintenance costs, including spare parts cost and depreciation, depending on the operation period of different categories of vehicles.

The environmental effect of using a toll road is estimated by environmental costs based on the intensity and structure of the traffic flow. According to the method (Gutarevych et al. 2006), the environmental costs (C_E, UAH billion) for overcoming the consequences of harmful substance emissions by vehicles during transportation are as follows:

$$C_E = c_E \cdot \sigma \cdot f \cdot G_\Sigma \cdot 10^{-9} \qquad (8.12)$$

where c_E is the damage caused to the environment by one conditional ton of pollutants, UAH/t; σ is the relative hazard coefficient, taking into account the pollution area type; f is the coefficient, taking into account scattering nature of gaseous substances and particles in the atmosphere depending on the substance deposition rate and natural conditions; and G_Σ is annual total emissions of pollutants by vehicles during road transportation, reduced to carbon monoxide, t. The annual total emissions G_Σ are calculated using dependency (8.13).

The annual total emissions of pollutants by vehicles during road transportation, reduced to carbon monoxide (Gutarevych et al. 2006) (G_Σ, t), are determined as follows:

$$G_\Sigma = \sum_{x=1}^{n} G_x \cdot R_x \qquad (8.13)$$

where G_x is annual emissions of the x-th harmful substance by vehicles during road transportation, t, which is determined using dependency (8.14), and R_x is the coefficient of relative aggressiveness for the x-th harmful substance (Gutarevych et al. 2006).

The annual emissions of the x-th harmful substance (G_x, t) by vehicles during road transportation are based on dependencies (8.2) and (8.4):

$$G_x = \frac{\sum_{i=1}^{6} \sum_{j=1}^{7} \sum_{k=1}^{11} \left(L_{Vijk} \cdot g_{ijkx}\right) \cdot \left(1 - (1 - k_{FUEL}) \cdot \frac{L_{road}^{road}}{L_{road}^{max}}\right)}{10^6}, \qquad (8.14)$$

where g_{ijkx} is specific emissions of the x-th harmful substance by vehicles of the i-th category, the j-th environmental class, using the k-th fuel type, g/km.

The total annual economic effect of the state (E_S, UAH billion) is based on dependencies (8.1), (8.6), (8.7), (8.9), (8.10), and (8.12). It is calculated as a result of the road construction project using Eq. 8.15:

$$E_S = \Delta T_\Sigma - \Delta C_E + \Delta T_A^{FUEL} + k_S \cdot (P_D - C_{constr\ i} - C_{maint\ i})$$
$$+ P_P \cdot (1 - k_S), \qquad (8.15)$$

where ΔT_Σ is the annual economic effect from tax amount increase coming to the state budget from enterprises of road transportation, which are carried out by the investigated road, UAH billion, which is calculated using dependencies (8.6) and (8.7); ΔC_E is the difference (effect) of the annual environmental costs for overcoming the consequences of harmful substance emissions by vehicles during transportation on the investigated road, UAH billion, which is calculated using dependency (8.12); ΔT_A^{FUEL} is the annual economic effect from the income increase to the state budget due to the excise tax on fuel sale for vehicles on the investigated road, UAH billion, which is calculated using dependency (8.9); and k_S is the state participation share in the project financing: $k_S = 1$ for 100% by the state financing, $k_S = 0.5$ for 50/50% by the state and private partner financing, and $k_S = 0$ for 100% by private partner financing.

The total annual economic effect of the private partner (E_P, UAH billion) is based on dependencies (8.1) and (8.10). It is calculated as a result of the road construction project using Eq. 8.16:

$$E_P = (P_D - C_{constr\ i} - P_P - C_{maint\ i}) \cdot (1 - k_S) \tag{8.16}$$

The payback period T_{PB} of the road construction project is defined as the number of years that have elapsed from the project beginning to the moment when the economic effect reaches a positive value:

$$T_{PB} = T_i \text{ for } \sum_{i=1}^{T_i} E_{P\ i} \text{ or } E_{S\ i} \geq 0, \tag{8.17}$$

where $E_{P\ i}$, $E_{S\ i}$ is the annual economic effect in the i-th year of the road construction or operation, UAH billion, based on dependencies (8.15) and (8.16).

Investment profitability is defined as the ratio of the economic effect E (determined using dependencies (8.15) and (8.16)) in the given year T_i of object operation after the payback period (8.17) to the invested funds value C_{constr} (8.1) indexed by the index i_I to the specific year T_i, %:

$$R = \frac{E}{C_{constr} \cdot i_1^{T_i - 1}} \cdot 100. \tag{8.18}$$

Thus, the developed mathematical model allows predicting the possible ecological and economic effect of the road construction project based on the public-private partnership. It also optimizes the road construction and operation parameters and financial relations between the state and private partner to achieve maximum effect by all project participants.

Using the proposed mathematical model, the algorithm and the MS Excel program to calculate the primary performance of road construction ecological and economic efficiency based on public-private partnership are developed.

8.4 Research Results

In the Kyiv region, due to geographical (presence of the Dnieper River, limited number of bridges across it, concentration of the bridges within the city of Kyiv) and economic features (significant concentration of business entities in Kyiv), traffic flows are directed toward Kyiv. Therefore, efficient transport connections in the Kyiv region are possible if there are modern highways with sufficient capacity around the city of Kyiv. It determines the topicality of the Great Ring Road construction in Kyiv. Such project requires significant financial investment and can be implemented on the basis of public-private partnership. At the same time, the implementation of this project may lead to a significant increase in harmful substance emissions by vehicles. Therefore, this study involves assessing the economic and environmental feasibility of the Great Ring Road construction project in Kyiv. In addition, it is necessary to determine the impact of specific parameters of the project implementation, particularly the state participation share in the project financing and the project traffic intensity, on the economic and environmental efficiency. Next, based on the analysis, it is necessary to establish the expedient values of these project parameters.

The ecological and economic efficiency of the Great Ring Road construction in Kyiv was evaluated using the developed mathematical model. This efficiency depends on the distribution of funding between public and private partners (k_S) and the project traffic intensity I_A (8.3).

The economic (E_S (8.15), E_P (8.16)) and environmental C_E (8.12) effects and their components are determined depending on the payback period T_{PB} (8.17) of spent funds by the state and the private partner and the 5 years of object effective operation after payback.

The most prolonged payback period T_{PB}^S (Fig. 8.2) is obtained for lower category roads (with lower design traffic intensity I_A).

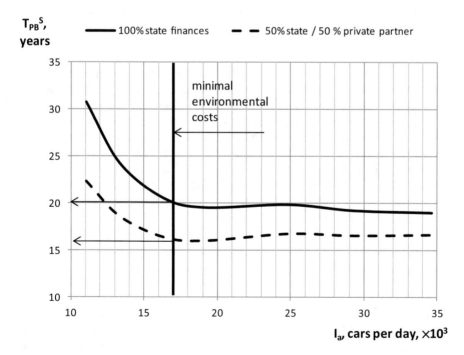

Fig. 8.2 Payback period of the Great Ring Road in Kyiv construction project based on public-private partnership for the state. (Adapted from Tsiuman 2020)

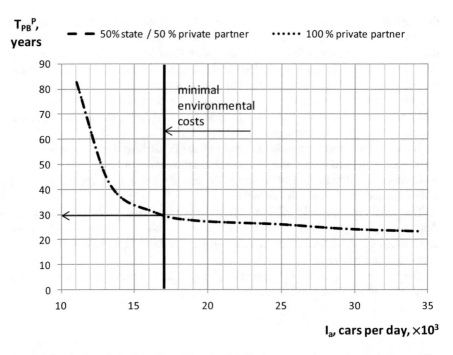

Fig. 8.3 Payback period of the Great Ring Road in Kyiv construction project based on public-private partnership for the private partner. (Adapted from Tsiuman 2020)

Also, the payback period of spent state resources T_{PB}^S depends on funding distribution between partners k_S. With the increase in project financing share for the private partner, the payback period of state resource costs T_{PB}^S decreases.

The payback of the private partner T_{PB}^P (Fig. 8.3) does not depend on the distribution of funding k_S because the ratio of costs and incomes of the private partner remains constant (Fig. 8.4). The total economic effect of the private partner E_P (8.16) is proportional to the state participation share k_S in the financing of construction (Fig. 8.4). It increases significantly for the traffic intensity I_A of $19 \cdot 10^3$ cars per day and more. This dependence is explained by the increase in road transport operation efficiency. As a result, it causes the increase in the traffic intensity I_A, reduction in the transportation operating costs C_{Vijk}^O (8.8), increase in the economic benefits for users, and, accordingly, increase in the incomes from driveway payment P_D (8.10).

With 100% by the private partner financing, he receives the total amount of income from driveway payment P_D^P (8.10) and pays the concession payment P_P (Fig. 8.9) based on the total object cost (equal to its construction cost) (Fig. 8.5). With 50% by the private partner financing, he receives half of the income from the driveway payment P_D^P (8.10) and pays the concession payment

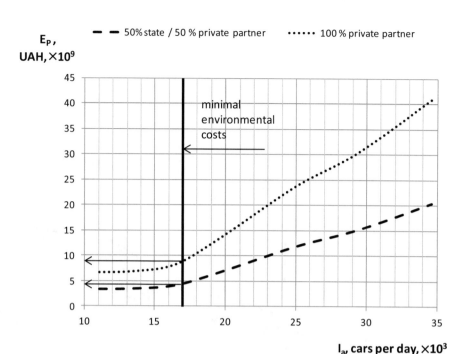

Fig. 8.4 Economic effect of the private partner from the implementation of the Great Ring Road in Kyiv construction project based on public-private partnership. (Adapted from Tsiuman 2020)

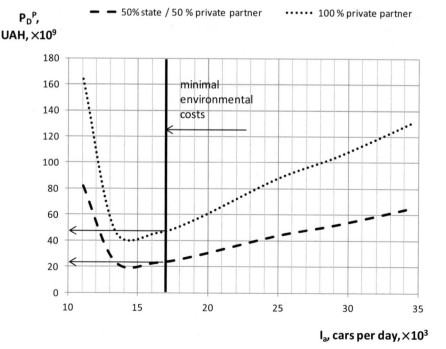

Fig. 8.5 Incomes from the driveway payment received by the private partner. (Adapted from Tsiuman 2020)

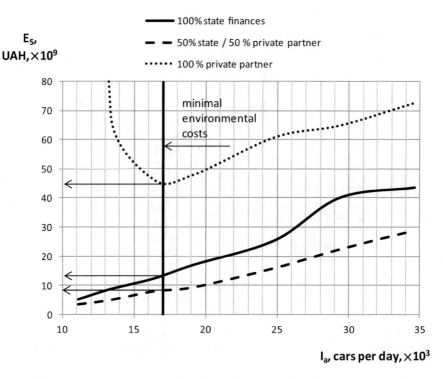

Fig. 8.6 Economic effect of the state from the implementation of the Great Ring Road in Kyiv construction project based on public-private partnership. (Adapted from Tsiuman 2020)

P_P based on half of the object cost. It is because the cost of the other half was paid by the state.

As a result, the economic effect of the state E_S (Fig. 8.6) and its components ΔT_{SUM} (Fig. 8.7), ΔC_E (Fig. 8.8), P_P (Fig. 8.9), and P_D^S (Fig. 8.10) are proportional to the payback period T_{PB}^S. The total economic effect of the state E_S (8.15) is also proportional to the state participation share in road construction financing k_S (Fig. 8.6). However, in the case of state participation in financing, growth of the total economic effect of the state E_S (Fig. 8.6) is more uniform than the economic effect of the private partner E_P (Fig. 8.4) if traffic intensity I_A increases. It is because the incomes increasing from driveway payment P_D^S (Fig. 8.10) is offset by the less rapid increasing of incomes from the excise tax on fuels ΔT_A^{FUEL} due to more efficient use of vehicles and increasing environmental cost ΔC_E (Fig. 8.8).

As a result of the road operation with forecasted transport and operational performance, incomes ΔT_{SUM} to the state budget from enterprise tax payments ΔT_Σ and fuel excise tax ΔT_A^{FUEL} are significantly increased (Fig. 8.7). At the same

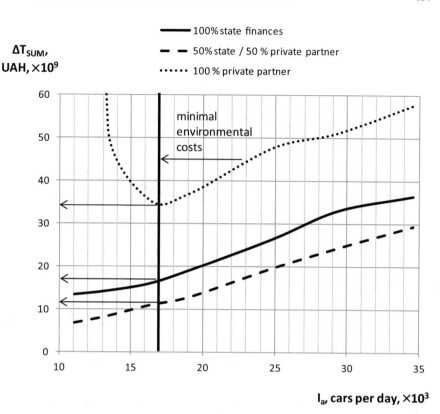

Fig. 8.7 Tax incomes from the implementation of the Great Ring Road in Kyiv construction project based on public-private partnership. (Adapted from Tsiuman 2020)

time, the amount of paid taxes ΔT_{SUM} increases intensively with the operation of better road.

Along with significant financial effects E_S, E_P, the designed road operation leads to a significant increase in environmental pollution by toxic emissions. According to the obtained values of annual emissions G_x, G_Σ by road transport, the effect of environmental costs increasing ΔC_E (8.12) from transportation by designed road was determined (Fig. 8.8).

Operation of the designed road will lead to increase in the environmental costs ΔC_E in comparison with operation of the existing roads due to a significant increase in traffic intensity I_A. These costs are about 1.5–7% of the total state economic effect E_S. The fuel excise tax ΔT_A^{FUEL} offsets these costs.

The total concession payment P_P, which is taken into account in the state effect E_S, is presented in Fig. 8.9.

The profitability of investments R_S (8.18) made by the state in the object implementation depends on its participation share k_S in financing. It increases with the

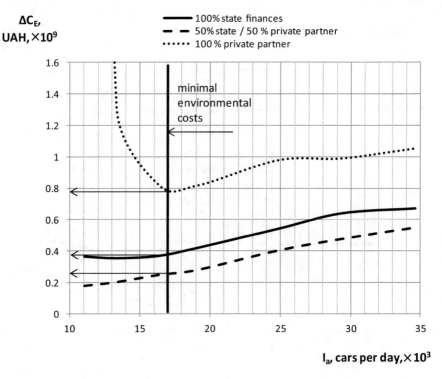

Fig. 8.8 Environmental costs from the implementation of the Great Ring Road in Kyiv construction project based on public-private partnership. (Adapted from Tsiuman 2020)

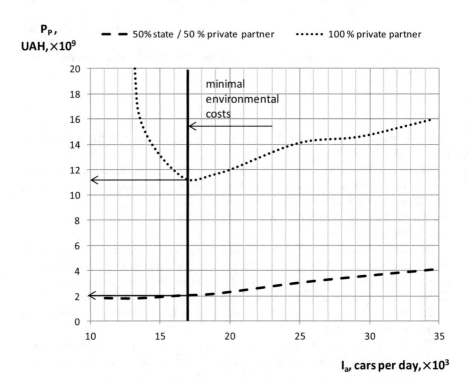

Fig. 8.9 Total concession payment from the implementation of the Great Ring Road in Kyiv construction project based on public-private partnership. (Adapted from Tsiuman 2020)

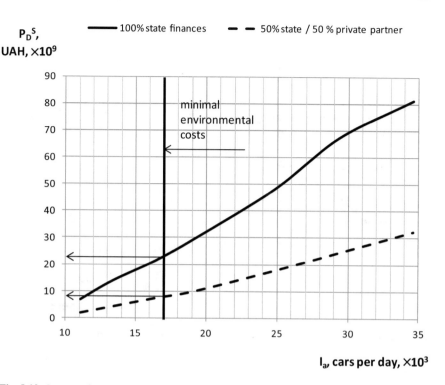

Fig. 8.10 Incomes from driveway payment received by the state. (Adapted from Tsiuman 2020)

private partner participation (Fig. 8.11). It is due to the concession payment P_P by the private partner to the state.

The investment profitability of private partner R_P (8.18) for the specified operation period does not depend on participation distribution of object financing k_S and is may be more than 25% with the successful project implementation (Fig. 8.12).

It is known that the average speed V_A and traffic uniformity are the most influential parameters on vehicle fuel consumption Q_{Vijk}^{FUEL}. According to the Great Ring Road construction project, the average speed V_A will increase by 3–3.2 times (from 25–35 to 75–110 km/h). Also, it is planned to increase the traffic uniformity (absence of traffic lights, intersections, and others). It will reduce fuel consumption Q_{Vijk}^{FUEL} to 30% (e.g., a car has 10 l per 100 km in the city and 7 l per 100 km on the highway). Such performance will correspond to the highest design traffic intensity I_A (road category). When the design intensity I_A decreases, the road transport efficiency will decrease. It will lead to a decrease in the economic effects E_S, E_P of the entire project.

Determination of the expedient traffic intensity I_A will ensure high economic efficiency E_S, E_P of construction and minimal environmental cost ΔC_E growth. In order to do this, the dependencies of the payback period T_{PB} (8.17) and environmental costs ΔC_E (8.12) on design traffic intensity I_A are compared (Fig. 8.13).

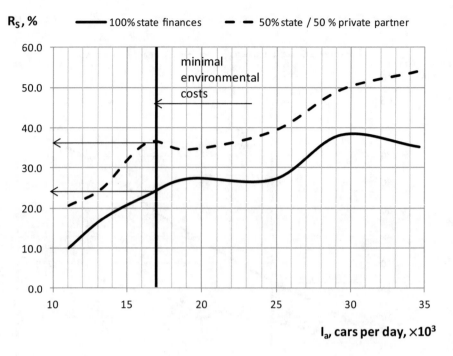

Fig. 8.11 Profitability of the state investments. (Adapted from Tsiuman 2020)

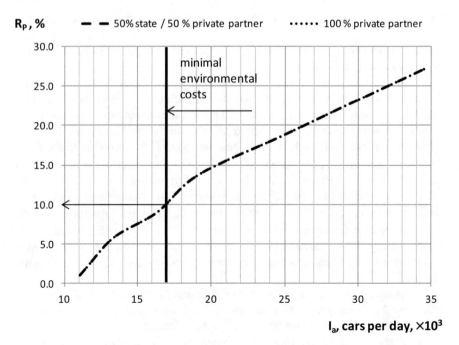

Fig. 8.12 Profitability of the private partner investment. (Adapted from Tsiuman 2020)

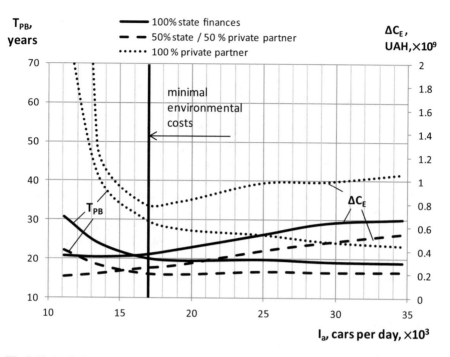

Fig. 8.13 Ecological and economic efficiency of road construction. (Adapted from Tsiuman 2020)

These dependencies show that as the intensity increases I_A, along with the payback period T_{PB} decrease of the state and private partner investments, the environmental costs ΔC_E will increase. It will mean an increase in harmful emissions G_x, G_Σ. At the same time, up to the traffic intensity I_A of $17 \cdot 10^3$ cars per day, there is a rapid decrease in the payback period T_{PB}. It slows down with a further increase in traffic intensity I_A. At the same time, the environmental costs ΔC_E begin to grow significantly with increase in traffic intensity I_A of more than $17 \cdot 10^3$ cars per day for various project financing options. In addition, in terms of project financing payback T_{PB} by the private partner for an intensity of $17 \cdot 10^3$ cars per day, there are minimal environmental costs ΔC_E. Therefore, the expedient traffic intensity I_A on the Great Ring Road will be of $17 \cdot 10^3$ cars per day according to specific environmental and economic performance of the construction. The payback period T_{PB} of the invested funds will be of 17...29 years.

8.5 Conclusions

The study allows us to draw the following conclusions:

1. Public-private partnership is one of the most effective forms of financial cooperation. It allows implementing of high-cost infrastructure projects, particularly

road construction. Built roads are operated on a toll basis. The key to the economic efficiency of the implemented project is the high traffic intensity. It guarantees a quick investment payback.

2. The increasing traffic intensity in implementing such road construction projects leads to a significant increase in harmful emissions. It must be taken into account when determining such project efficiency. To this end, the model for evaluating road construction's environmental and economic efficiency based on public-private partnership has been developed. The model allows assessing the economic and environmental effects of project participants. It also allows determining tariffs for road use, taking into account the effect of reducing the operating costs by vehicle owners.

3. Using the developed model, the ecological and economic effects from the realization of the Great Ring Road in the Kyiv construction project are estimated. According to the estimation results, the expedient traffic intensity on the designed road has been established. The expedient traffic intensity ensures the minimum payback period of spent funds for construction. Also, it ensures the minimum growth of harmful emissions by traffic.

References

Bazyliuk A, Zhulyn O (2007) State regulation and control over tariff policy in the market of toll roads. Proc Natl Trans Univ 15:168–174

Gutarevych Y et al (2006) Ecology and automobile transport: textbook. Aristei, Kyiv

Kanylo P (2013) Automobile transport. Fuel and environmental problems and prospects. KhNAHU, Kharkiv

Mateichyk V (2006) Evaluation methods and methods for improving environmental safety of road vehicles. NTU, Kyiv

Tsiuman YS (2020) Economic mechanism of road enterprises investment activity activation on public-private partnership basis. Dissertation, National Transport University

Varnavskyi V (2011) Public-private partnership: some issues of theory and practice. World Econ Int Relat 9:41–50

Vdovenko YS (2008) Private-state sources of financing for the development of roads in the region. Dissertation, Chernihiv State Technological University

Zapatrina I (2010) Public-private partnership in Ukraine: application prospects for infrastructure projects implementation and provision of public services. Econ Forecasting 4:62–86

Zhulyn OV (2009) Tariffing of services for a driveway on toll roads of Ukraine. Dissertation, National Transport University

Chapter 9
Improving the Energy Efficiency and Environmental Performance of Vehicular Engine Equipped Within the On-Board Hydrogen Production System

Mykola Tsiuman, Vasyl Mateichyk, Miroslaw Smieszek, Ivan Sadovnyk, Roman Artemenko, and Yevheniia Tsiuman

Nomenclature

ICE	Internal combustion engine
HCG	Hydrogen-containing gas
H_2	Hydrogen
O_2	Oxygen
EG	Exhaust gas
TEG	Thermoelectric generator
VW BBY	Volkswagen engine of BBY model
OBD	On-board diagnostic
ECU	Electronic control unit
UNECE	The United Nations Economic Commission for Europe
CO	Carbon monoxide
CO_2	Carbon dioxide
C_mH_n	Hydrocarbons
NO_x	Nitrogen oxides

M. Tsiuman (✉)
Faculty of Automotive and Mechanical Engineering, National Transport University, Kyiv, Ukraine
e-mail: tsumanmp@ntu.edu.ua

© The Author(s), under exclusive license to Springer Nature Switzerland AG 2023
S. Boichenko et al. (eds.), *Sustainable Transport and Environmental Safety in Aviation*, Sustainable Aviation, https://doi.org/10.1007/978-3-031-34350-6_9

9.1 Introduction

Modern vehicles mainly use an internal combustion engine (ICE) as a propulsion system. Such engines that ensure their high environmental safety are equipped with exhaust gas (EG) purification systems with catalytic converters (Gritsuk et al. 2017, 2018; Tsiuman 2019). At the same time, toxic combustion products are converted in the catalytic converter into nontoxic carbon dioxide and water, which leaves open the problem of ensuring carbon neutrality of automobile engine emissions. Reducing the consumption of carbon fuels allows us to approach the solution of this problem.

The most appropriate in terms of reducing the consumption of carbon fuels today is the use of hydrogen. Therefore, a lot of scientific research papers contain the description of the study of such substitution since the middle of the last century. However, complete replacement requires the design of new types of propulsion systems, which does not allow to quickly replace a large number of existing vehicles. In addition, the problem of complete replacement of carbon fuels with hydrogen is complicated by the lack of hydrogen storage and the lack of mobile hydrogen generators of sufficient capacity. An effective alternative to complete hydrogen replacement is the hydrogen-containing gas (HCG) used as additive to the fresh charge of automobile engines.

The most appropriate is the use of HCG produced by electrolysis of a water solution of alkaline. This is due to the sufficient simplicity of the design of the electrolyzer and the lack of need to store large amounts of explosive gas on board the vehicle. This method application provides an additive to the fresh engine charge of a certain amount of HCG, which contains gases of oxygen (O_2) and hydrogen (H_2). Numerous studies contain the description of the use of such HCG on various types of ICEs (Madyira and Harding 2014; Le et al. 2013; Gutarevych et al. 2018; D'Andrea et al. 2003; Wang et al. 2011, 2012; Lakshmi et al. 2013; Bari and Mohammad-Esmaeil 2010; Dandrea 2004). According to the results of these studies, the addition of a small amount of HCG to the fresh charge improves fuel efficiency and energy performance of the engine at low loads due to the reduction of combustion time. In addition, there is a decrease of carbon monoxide, hydrocarbon, and nitrogen oxide emissions. However, a significant disadvantage of using electrolysis to obtain HCG on board the vehicle is the additional energy consumption for gas production, which is proportional to the amount added to the fresh engine charge. This shortcoming can be eliminated by recovering the thermal energy of ICE, which is lost in the cooling or exhaust systems.

One of the possible options for the recuperation of lost heat energy of ICEs is the use of thermoelectric conversion. A lot of scientific papers describe the study of the use of thermoelectric generators (TEG) on ICEs (Anatychuk and Kuz 2012; Kuz 2012; Gao and Rowe 2002; Fairbanks 2013; Vijayagopal and Rousseau 2015; Zheng and Fan 2016; Dmytrychenko et al. 2018). At the same time, the issue of the possibility of using TEG to ensure the operation of the on-board hydrogen production system has not been previously studied. This study was performed using the results of previous work of the authors (Tsiuman and Artemenko 2016;

9 Improving the Energy Efficiency and Environmental Performance...   145

Artemenko and Tsiuman 2016; Tsiuman et al. 2019, 2020; Mateichyk et al. 2020; Kuric et al. 2018; Dyachenko 2008), which present some results of evaluating the efficiency of HCG on ICE and conversion of EG thermal energy into TEG, equipment, and software for studying ICE performance when using HCG with TEG, mathematical models of fuel efficiency and environmental performance of the engines and vehicles. New in this study is to define the appropriate parameters of the on-board hydrogen production system with TEG to ensure improvement of the energy efficiency and environmental performance of vehicular engine.

9.2 Hydrogen Production System

The scheme of vehicular engine equipped within the on-board hydrogen production system using the recovery of EG thermal energy in TEG is shown in Fig. 9.1.

Addition of HCG to an air charge is carried out during operation of the engine in the modes of low load and idling. The required amount of HCG is produced by the hydrogen generator, which adds the obtained hydrogen to the engine air charge after the air filter. Ensuring the operation of the HCG generator is carried out by power from an additional battery.

Electric energy for charging the additional battery is obtained during the medium and high engine load operation mode due to the recovery of EG thermal energy in the TEG. Thus, the additional battery provides the necessary supply of electricity for the hydrogen production system operation during the vehicular engine operation. TEG

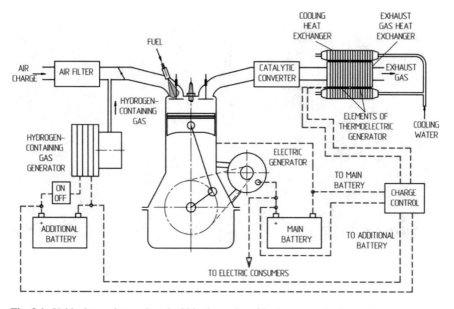

Fig. 9.1 Vehicular engine equipped within the on-board hydrogen production system

is installed in the exhaust pipeline as next one after the catalytic converter. This installation provides sufficient efficiency of purification of EG from harmful substances. The required temperature difference for thermoelectric conversion is provided by a system of heat exchangers for EG and coolant.

9.3 Experimental Procedure

Experimental tests of the vehicular engine equipped within the on-board hydrogen production system using the recovery of EG thermal energy in TEG were performed in the laboratory of engine tests in National Transport University. These tests were performed on a VW BBY engine (Fig. 9.2) used on a Skoda Fabia. A technical description of main parameters of the object of experimental research has been presented in Table 9.1.

The main technical parameters of the on-board hydrogen production system using the recovery of EG thermal energy in TEG, which is designed for experimental tests, are presented in Table 9.2. Serial and parallel circuits were used for electrical connection of thermoelectric modules of TEG. This allowed to obtain an electrical voltage under load at the level of 14 V.

The working load on the engine during the tests was created by an electric brake machine GPF a17h with a power of 250 kW, mounted on the bench of the model SAK-670 (Fig. 9.3). The maximum frequency of the bench rotor is 3200 rpm. Therefore, low-speed gear of mechanical gearbox was used to create higher engine speeds.

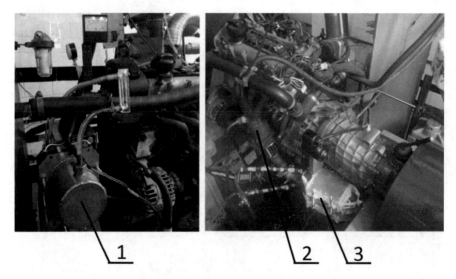

Fig. 9.2 The tested engine within the researched system, which includes the next ones: hydrogen generator (1), catalytic converter of EG cleaning system (2), and TEG (3)

Table 9.1 Brief technical description of the VW BBY engine used on Skoda Fabia

Parameter	Specification
Vehicle weight, kg	1100
Fuel	Petrol
Number of cylinders/arrangement of cylinders	4/inline
Displacement, l	1.39
Diameter/stroke, mm	76.5/75.6
Compression ratio	10.5
Inlet/exhaust valves amount per cylinder	2/2
Torque, N·m/speed, rpm	126/3800
Power, kW/speed, rpm	55/5000
EG cleaning system	Three-component catalytic converter
Ratios of gearbox	3.455 2.095 1.387 1.025 0.813
Main gear ratio	3.882
Rolling radius of wheel, m	0.265
Maximum speed, km/h	167
Vehicle electric system voltage, V/current, A	14/70

Table 9.2 Main technical parameters of the on-board hydrogen production system using the recovery of EG thermal energy in TEG

Parameter	Specification
HCG producing method	Electrolysis
Working resource for the hydrogen generator	KOH aqueous solution
Output of hydrogen, l/min	3
Power of HCG production, W	360
Additional battery capacity, A·h	30
TEG heat exchangers material	Aluminum
TEG element power, W	19
TEG maximum temperature, °C	400
TEG voltage, V	14
Number of TEG elements	8
TEG efficiency	0.035

The experimental and measuring setup includes (Fig. 9.3) 1. VW BBY engine; 2. test bench with an electric braking device; 3. device to measure torque and crankshaft rotation speed; 4. device to measure fuel consumption; 5. devices to measure C_{CO}, $C_{C_mH_n}$, C_{NO_x}, and C_{CO_2} EG concentrations; 6. in-cylinder pressure sensor; 7. analog-to-digital converter; 8. OBD II adapter to connect to engine control

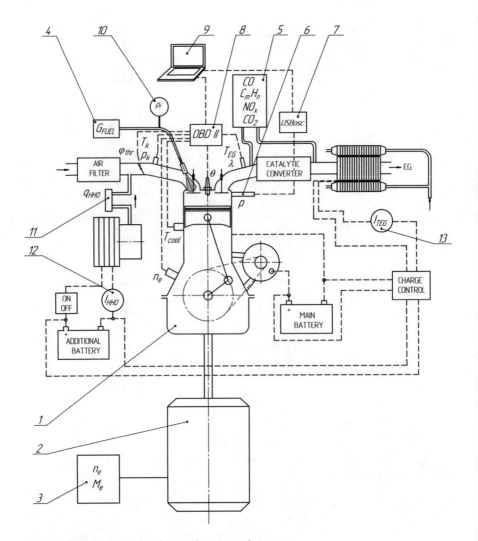

Fig. 9.3 The experimental and measuring setup scheme

system diagnostic line; 9. personal computer; 10. device to control fuel system pressure; 11. device to measure hydrogen-containing gas generator productivity; 12. device to measure electric current in hydrogen-containing gas generator power supply circuit; and 13. device to measure electric current in additional battery charge supply circuit from TEG.

During the tests of the engine, its parameters were determined at speeds of 1200, 2100, 2400, 3000, and 3800 rpm when changing the load from minimum to maximum, with the addition of HCG and without its addition. The fuel-air mixture composition was maintained in stoichiometric by feedback on the signals of the lambda sensor to ensure high efficiency of the catalytic converter. Therefore, the

9 Improving the Energy Efficiency and Environmental Performance... 149

addition of HCG did not cause any changes in the composition of the combustible mixture relative to the stoichiometric composition and did not reduce the efficiency of EG purification.

Experimental tests allowed to determine the following parameters: engine speed n_e, torque M_e, coefficient of air excess λ, fuel consumption G_{FUEL}, angle of ignition timing θ, angle of throttle position φ_{thr}, absolute inlet air pressure p_k and temperature T_k, in-cylinder pressure p, coolant temperature T_{COOL}, EG temperature T_{EG}, concentrations of carbon dioxide C_{CO_2}, nitrogen oxides C_{NO_x}, total hydrocarbons $C_{C_mH_n}$ and carbon monoxide C_{CO} in EG till catalytic converter, and, behind it, electric current in supply circuit of the HCG generator I_g.

According to the measured parameters, the following ones were determined: engine power N_e, air consumption G_{AIR}, specific fuel consumption g_e, indicated engine power N_i, volumetric engine efficiency η_v, mechanical loss power N_m, in-cylinder maximum pressure p_{max} and temperature T_{max}, crankshaft position at the rapid combustion beginning φ_{RC}, post-combustion beginning $\varphi_{p_{max}}$ and combustion end $\varphi_{x_{max}}$, speed of pressure increase during combustion $\Delta p/\Delta\varphi$, Vibe function parameters (Dyachenko 2008): heat emission character m and combustion time φ_Z, CO, C_mH_n, NO_x and CO_2 emissions, catalytic converter efficiency E_{CO}, $E_{C_mH_n}$, E_{NO_x}.

The determination of the working body temperature in the cylinder during combustion is based on the current composition of the working body, which contains air, fuel and combustion products, gas constants of these components, the current values of pressure, and volume of the working body.

During the tests, individual engine parameters were determined using the OBD-II standard diagnostic system and software and hardware, which is described in detail in (Kuric et al. 2018). Emissions of harmful substances from EG were determined using the measured concentrations of CO, C_mH_n, CO_2 (infrared radiation selective absorption method), and NO_x (chemical luminescence method).

A set of software and hardware was used to analyze the engine in-cylinder operating process (Mateichyk et al. 2020). This equipment included, in particular, pressure sensor (strain gauge principle) installed in the engine cylinder head, amplifier of pressure sensor voltage, analog-to-digital converter (ADC) to submit voltage data of ignition time, crankshaft rotation speed and in-cylinder pressure on computer, and software to indicate, process, and analyze the obtained data of engine operating process (Fig. 9.4).

9.4 Mathematical Procedure

A mathematical model is used to study the effect of using the developed on-board hydrogen production system using the recovery of EG thermal energy in TEG on fuel economy and harmful emissions by the vehicular engine while vehicle motion is in the driving cycle. The engine experimental tests results allowed to refine this

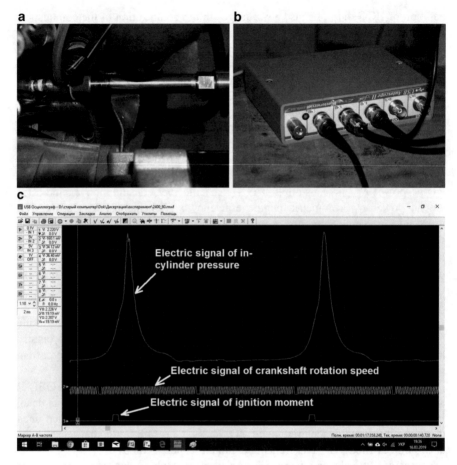

Fig. 9.4 Pressure sensor (**a**), ADC (**b**), and example of data indication at computer program (**c**)

model in the part of simulation of the processes of intake, compression, and combustion with the addition of HCG.

The mathematical modeling uses the method of calculating the operating process in the engine cylinder on the principle of volume balance, the method of mathematical description of the combustion process with Vibe equation, and the methods of mechanism theory for modeling friction losses in kinematic pairs of engine (Dyachenko 2008). Some aspects of modeling the processes of coolant and catalytic reactor heating, which have a significant impact on the engine fuel efficiency and harmful emissions, are described in detail in (Gritsuk et al. 2017, 2018).

Simulation of the engine effective parameters with the on-board hydrogen production system is carried out according to the algorithm presented in Fig. 9.5. The algorithm describes the main processes that occur in the engine, catalytic converter,

9 Improving the Energy Efficiency and Environmental Performance... 151

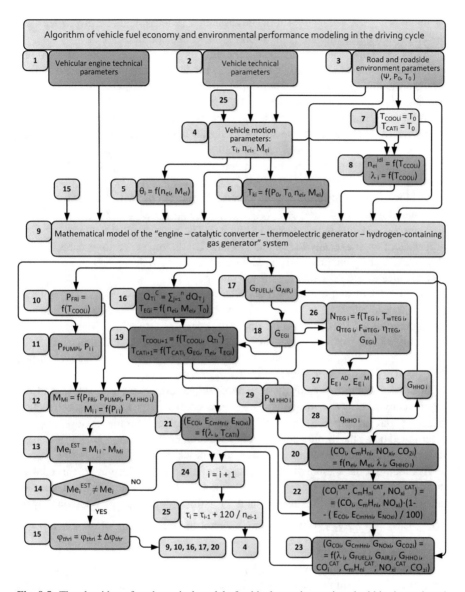

Fig. 9.5 The algorithm of mathematical model of vehicular engine equipped within the on-board hydrogen production system using the recovery of EG thermal energy in TEG

TEG, and HCG generator. The main stages of the calculation are presented in the appropriate blocks.

Blocks 1–3 contain the input data required for simulation.

In blocks 1 and 2, the engine and vehicle parameters are defined. The numerical values of these parameters are detailed in Table 9.1.

In block 3, such parameters of the environment is determined: air pressure p_0, Pa, the ambient temperature T_0, K, and rolling resistance coefficient, ψ.

Initial data for the mathematical modeling is prepared in blocks 4–8 of the algorithm. In block 4, the engine operating mode parameters at the current time are determined. They are time from starting of the engine τ_i, s; current engine speed $n_{e\ i}$, min^{-1}; and current engine torque $M_{e\ i}$, N·m.

In block 5, ignition timing angle θ_i is determined. It is presented by a function $\theta_i = f(n_{e\ i}, M_{e\ i})$. In block 6, the engine intake air temperature is determined using a function $T_{k\ i} = f(p_0, T_0, n_{e\ i}, M_{e\ i})$, K. The initial values of engine coolant temperature $T_{\text{COOL}\ i} = T_0$, K, and catalytic converter temperature $T_{\text{CAT}\ i} = T_0$, K, are determined in block 7.

In block 8, current idling engine speed $n_{e\ i}^{idl}$ and coefficient of air excess λ_i are determined using a function of $T_{\text{COOL}\ i}$: $n_{e\ i}^{idl} = f(T_{\text{COOL}\ i}), \lambda_i = f(T_{\text{COOL}\ i})$.

The prepared data is transferred to block 9. The mathematical modeling of main processes of engine, catalytic reactor, TEG, and hydrogen generator is carried out in this block. During the processes, modeling the energy efficiency and environmental performance of the vehicular engine are determined.

In block 10, mean pressure of friction loss in the engine $p_{FR\ i} = f(T_{\text{COOL}\ i})$, Pa, is determined. In block 11, mean pressure of pumping loss $p_{PUMP\ i}$, Pa, and indicated mean pressure $p_{i\ i}$, Pa, of the engine operating cycle are determined. The mechanical loss moment $M_{M\ i} = f(p_{FR\ i}, p_{PUMP\ i}, p_{M\ HHO\ i})$, N·m, and engine indicated torque $M_{i\ i} = f(p_{i\ i})$, N·m, are determined in block 12. In block 13, effective torque, $M_{e\ i}^{EST} = M_{i\ i} - M_{M\ i}$, N·m, is determined.

In block 14, the values of the estimated engine effective torque $M_{e\ i}^{EST}$ and the required engine effective torque $M_{e\ i}$ are compared. If $M_{e\ i}^{EST} \neq M_{e\ i}$, then a calculation is carried out in block 15. In block 15, the value of the throttle position $\varphi_{thr\ i}$ is determined using the throttle position adjustment value $\Delta\varphi_{thr\ i}$: $\varphi_{thr\ i} = \varphi_{thr\ i} \pm \Delta\varphi_{thr\ i}$. If $M_{e\ i}^{EST} = M_{e\ i}$, then modeling is continued in block 24.

In block 16, heat amount transferred from the in-cylinder gas to the coolant during the operating cycle $Q_{T\ i}^C$ is determined. It is determined as a sum of heat amount removed into the coolant in estimated periods of operating cycle $dQ_{T\ j}$: $Q_{T\ i}^C = \sum_{j=1}^{n} dQ_{T\ j}$, J. It takes into account the temperature of the exhaust gas $T_{EG\ i} = f(n_{e\ i}, M_{e\ i}, T_0)$, K.

In block 17, fuel consumption $G_{FUEL\ i}$, kg/h, and air consumption $G_{AIR\ i}$, kg/h, by the engine are determined. Then, in block 18, mass flow of exhaust gas, $G_{EG\ i}$, kg/h, is determined. In block 19, the coolant temperature $T_{\text{COOL}\ i+1} = f\left(T_{\text{COOL}\ i}, Q_{T\ i}^C\right)$, °C, and the catalytic converter temperature $T_{\text{CAT}\ i+1} = f(T_{\text{CAT}\ i}, G_{EG\ i}, n_{e\ i}, T_{EG\ i})$, K, are determined.

In blocks 20–23, the environmental performance of vehicular engine is determined. The concentrations of CO (%), C_mH_n (ppm), NO_x (ppm), and CO_2 (%) till the catalytic converter, $(CO_i, C_mH_n\ _i, NO_x\ _i, CO_2\ _i) = f(n_{e\ i}, M_{e\ i}, \lambda_i, G_{HHO\ i})$, are determined in block 20. In block 21, the efficiency of cleaning from CO, C_mH_n, NO_x, %, namely, $(E_{CO\ i}, E_{CH\ i}, E_{NOx\ i}) = f(\lambda_i, T_{\text{CAT}\ i})$, is determined. In block 22, the

9 Improving the Energy Efficiency and Environmental Performance... 153

concentrations of CO (%), C_mH_n (ppm), NO_x (ppm) behind the catalytic converter, namely, $\left(CO_i^{CAT}, C_mH_{n_i}^{CAT}, NO_{x_i}^{CAT}\right) = (CO_i, C_mH_{n\ i}, NO_{x\ i}) \cdot (1 - (E_{CO\ i},$ $,E_{CH\ i}E_{NOx\ i})/100)$, are determined. In addition, in block 23, the mass emissions of CO, C_mH_n, NO_x, and CO_2 (kg/h), namely,

$$\left(G_{CO\ i}, G_{C_mH_n\ i}, G_{NO_x\ i}, G_{CO2\ i}\right)$$
$$= f\left(\lambda_i, G_{FUEL\ i}, G_{AIR\ i}, G_{HHO\ i}; \left(CO_i^{CAT}, C_mH_{n_i}^{CAT}, NO_{x_i}^{CAT}, CO_{2\ i}\right)\right),$$

are determined.

Current power of the thermoelectric generator $N_{TEG\ i}$, (W), is determined in block 26 as $N_{TEG\ i} = f(T_{EG\ i}, T_{wTEG\ i}, q_{TEG\ i}, F_{wTEG}, \eta_{TEG}, G_{EG\ i})$. It depends on the following parameters: heat flow density $q_{TEG\ i}$ through EG heat exchanger walls, current temperature of the heat exchanger walls $T_{wTEG\ i}$, EG temperature $T_{EG\ i}$, heat exchanger walls area F_{wTEG}, thermoelectric generator efficiency η_{TEG}, EG mass flow $G_{EG\ i}$.

Block 27 determines the current value of accumulated electric energy in main $E_{E\ i}^M$ and additional $E_{E\ i}^{AD}$ batteries depending on current power of the thermoelectric generator. Part of this energy is used to provide hydrogen-containing gas generator productivity $q_{HHO\ i}$. The last one is determined in block 28. If the electric energy in additional battery is not enough to power the hydrogen-containing gas generator, then last one is powered by main battery. This leads to consume additional mechanical energy by electric generator and creates additional mechanical loss of engine. Block 29 determines mean pressure of mechanical loss on hydrogen-containing gas production $p_{M\ HHO\ i}$. Block 30 determines hydrogen-containing gas consumption per hour $G_{HHO\ i}$ that is mixed with air charge of the engine. The hydrogen-containing gas consumption depends on hydrogen-containing gas generator productivity.

After defining the energy efficiency and environmental performance of the vehicular engine, in block 24, the calculation cycle number is changed to the next one, $i = i + 1$. Current time from the engine starting is changed in block 25: $\tau_i = \tau_{i-1} + 120/n_{e\ i-1}$, s.

To determine the amount of fresh charge, which consists of fuel, air, and HCG, the mathematical model uses a differential approach. This allows you to determine the current mass of fresh charge entering the intake into the cylinder. The total amount of fresh charge (M_{FC}, kg) is determined as the current values sum in the calculation periods of the intake process using the following equation:

$$M_{FC} = \int_{\varphi_{op\ in}}^{\varphi_{cl\ in}} w_{FC\ i} \cdot \mu f_{in\ i} \cdot \frac{p_{FC\ i}}{T_{FC\ i} \cdot R_{FC}} d\tau, \qquad (9.1)$$

where $\varphi_{op\ in}$ is the crankshaft position of opening of the inlet valve, degrees; $\varphi_{cl\ in}$ is the crankshaft position of closing of the inlet valve, degrees; $w_{FC\ i}$ is the current speed of flow of fresh charge in cross section of the inlet valve, m/s; $\mu f_{in\ i}$ is the current effective inlet valve cross section, m^2; $p_{FC\,i}$ is the current pressure of fresh charge in front of the inlet valve, Pa; $T_{FC\,i}$ is the current temperature of fresh charge in front of the inlet valve, K; R_{FC} is a specific fresh charge gas constant, J/(kg·K); and $d\tau$ is period of calculating time, s.

The specific fresh charge gas constant (R_{FC}, J/(kg·K)) is defined with the following equation:

$$R_{FC} = \frac{R_{HHO} \cdot \lambda \cdot l_0 \cdot \frac{g_{HHO}}{100} + \left(1 - \frac{g_{HHO}}{100}\right) \cdot \left(\frac{8314}{\mu_{FUEL}} + R_{AIR} \cdot \lambda \cdot l_0\right)}{1 - \frac{g_{HHO}}{100} + \lambda \cdot l_0}, \qquad (9.2)$$

where R_{HHO} is a specific HCG gas constant, J/(kg·K); R_{AIR} is a specific gas constant for air, J/(kg·K); λ is a coefficient of air excess; l_0 is a stoichiometric air requirement, kg/kg; g_{HHO} is the mass fraction of HCG in the fresh charge, %; and μ_{FUEL} is a molar mass of fuel, kg/kmol.

The fraction of HCG in the fresh charge depends on the performance of the hydrogen generator and the engine operation mode. This is determined by the amount of air consumed by the engine per cycle. This fraction was determined during experimental tests of the engine and presented in the mathematical model as a dependence (g_{HHO}, %):

$$\begin{aligned}
g_{HHO} = \ &2.113 - 1.10852 \cdot 10^{-3} \cdot n_e - 6.368 \cdot 10^{-2} \cdot M_e \\
&+ 1.023 \cdot 10^{-7} \cdot n_e^{\,2} + 8.094 \cdot 10^{-4} \cdot M_e^{\,2} + 3.446 \cdot 10^{-5} \cdot n_e \cdot M_e \\
&+ 4.592 \cdot 10^{-11} \cdot n_e^{\,3} - 4.844 \cdot 10^{-6} \cdot M_e^{\,3} - 6.699 \cdot 10^{-9} \cdot n_e^{\,2} \cdot M_e \\
&- 2.847 \cdot 10^{-7} \cdot n_e \cdot M_e^{\,2} - 8.139 \cdot 10^{-15} \cdot n_e^{\,4} + 1.068 \cdot 10^{-8} \cdot M_e^{\,4} \\
&+ 8.508 \cdot 10^{-10} \cdot n_e \cdot M_e^{\,3} + 2.558 \cdot 10^{-11} \cdot n_e^{\,2} \cdot M_e^{\,2} \\
&+ 4.55 \cdot 10^{-13} \cdot n_e^{\,3} \cdot M_e, \qquad (9.3)
\end{aligned}$$

where M_e is an engine torque, N·m.

Then, HCG quantity in the fresh charge (G_{HHO}, kg) is defined with the following equation:

$$G_{HHO} = \frac{M_{FC} \cdot \frac{g_{HHO}}{100} \cdot \lambda \cdot l_0}{1 - \frac{g_{HHO}}{100} + \lambda \cdot l_0}. \qquad (9.4)$$

The hydrogen generator productivity providing the quantity of HCG G_{HHO} production (q_{HHO}, l/min) is described by the following equation:

$$q_{HHO} = \frac{G_{HHO} \cdot i_c \cdot n_e \cdot R_{HHO} \cdot T_0 \cdot 1000}{2 \cdot p_0}, \qquad (9.5)$$

where i_c is a number of cylinders of engine; n_e is engine speed, rpm; p_0 is an ambient air pressure, Pa; and T_0 is an ambient air temperature, K.

9 Improving the Energy Efficiency and Environmental Performance. . .

The hydrogen addition influences the combustion. The combustion time and character are taken into account during the modeling with Vibe equation (Dyachenko 2008). A current burned fuel fraction (x_i) is described as

$$x_i = 1 - e^{-6.908 \cdot \left(\frac{\varphi_i - (360 - \theta)}{\varphi_z \left(n_e, M_e, g_{HHO} \right)} \right)^{m \left(n_e, M_e, g_{HHO} \right) + 1}}, \qquad (9.6)$$

where φ_i is the current crankshaft position, degrees; θ is timing angle of ignition, deg.; $\varphi_z(n_e, M_e, g_{HHO})$ is combustion time as a function of speed and load of the engine and hydrogen addition, degrees; and $m(n_e, M_e, g_{HHO})$ is combustion character as a function of speed and load of the engine and hydrogen addition.

Because the hydrogen-containing gas production may wait electric energy consumption from main battery, additional electric generator load, and additional mechanical loss, then this takes into account in determining engine mechanical loss. The mean pressure of mechanical loss on hydrogen-containing gas production ($p_{M\ HHO}$, Pa) is described by the following equation:

$$p_{M\ HHO} = \frac{30 \cdot \tau \cdot n_{G\ HHO} \cdot q_{HHO}}{i_c \cdot V_h \cdot n_e \cdot \eta_{TEG}}, \qquad (9.7)$$

where τ is number of strokes in engine working cycle; $n_{G\ HHO}$ is HCG generator consumed electric power, W/(l/min); i_c is number of engine cylinders; V_h is swept volume of cylinder, m^3; and η_{TEG} is efficiency coefficient of conversion of energy by TEG.

The power of the thermoelectric generator ($N_{TEG\ i}$, W) is described by the following equation:

$$N_{TEG} = \frac{Q_{TEG} \cdot i_c \cdot n_e \cdot \eta_{TEG}}{30 \cdot \tau}, \qquad (9.8)$$

where Q_{TEG} is heat quantity that is given back from EG to heat exchanger walls during exhaust process of working cycle, J.

The heat quantity that is given back from EG (Q_{TEG}, J) is determined by the following equation:

$$Q_{TEG} = \int_{\varphi_{op\ ex}}^{\varphi_{cl\ ex}} [\alpha_{wTEG\ i} \cdot (T_{EG\ i} - T_{wTEG\ i}) \cdot F_{wTEG}] d\tau, \qquad (9.9)$$

where $\varphi_{op\ ex}$ is the crankshaft position of opening of the exhaust valve, degrees; $\varphi_{cl\ in}$ is the crankshaft position of closing of the exhaust valve, degrees; $\alpha_{wTEG\ i}$ is current heat output coefficient of EG in the heat exchanger walls, W/(m^2·K); $T_{EG\ i}$ is current EG temperature in heat exchanger, K; $T_{wTEG\ i}$ is current temperature of the EG heat exchanger wall in TEG, K; and F_{wTEG} is working surface area of the EG heat exchanger walls in TEG, m^2.

The current electrical energy value that accumulated in the additional battery ($E^{AD}_{E\ i}$, J) is determined by the following equation:

$$E^{AD}_{E\ i} = E^{AD}_{E\ 0} + N_{TEG} \cdot d\tau - n_{G\ HHO} \cdot q_{HHO} \cdot d\tau, \quad (9.10)$$

where $E^{AD}_{E\ 0}$ is an initial electrical energy value in the additional battery, J.

9.5 Results and Discussion

The use of the developed method of experimental tests of the engine and mathematical modeling allowed to study the effect of HCG adding to fresh charge on operating process performance, fuel consumption, emissions of engine pollutants in various modes, and the vehicle environmental performance and fuel efficiency during its conditional motion in the driving cycle using TEG as a source of additional energy.

The experimental dependencies of hourly and specific fuel consumption on the engine speed and load with different addition of HCG are shown in Fig. 9.6.

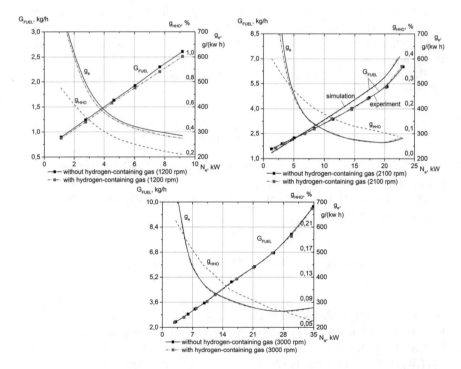

Fig. 9.6 Dependencies of hourly and specific fuel consumption on the engine speed and load with different addition of HCG

9 Improving the Energy Efficiency and Environmental Performance...

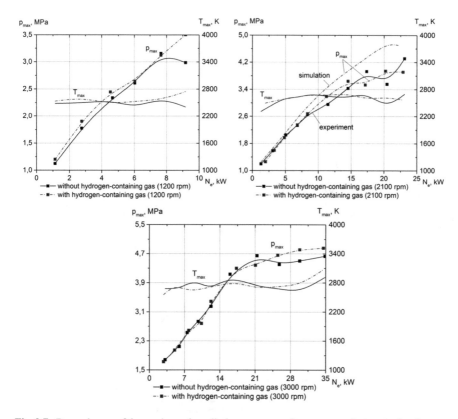

Fig. 9.7 Dependences of the maximum in-cylinder pressure and temperature during combustion on the engine speed and load with different HCG addition

The presented dependences show that the greatest influence of HCG adding to the air charge on consumption of fuel is manifested at low speeds and loads of engine. For example, at the engine mode with speed of 1200 rpm and load of up to 1 kW, the HCG mass fraction in the fresh charge of 0.75% reduces fuel consumption by 2.4%, and at the engine mode with load of 9 kW, the HCG fraction of 0.2% reduces fuel consumption by 3.8%. At the engine mode with speed of 2100 rpm, the HCG addition is 0.1... 0.4%, which at a load of 2 kW reduces fuel consumption by 3.5%. At higher loads and speeds, fuel consumption is virtually independent of the addition of HCG.

To study the reasons for the reduction of fuel consumption with HCG addition to the engine fresh charge, the parameters of its operating process were studied. As you know, the main operating process parameters at the internal combustion engine are the values of the maximum in-cylinder pressure and temperature during combustion. The dependences of these parameters on the engine speed and load are presented in Fig. 9.7. These dependences confirm that in those operating modes of the engine, where there was a reduction in fuel consumption due to the addition of HCG, there is

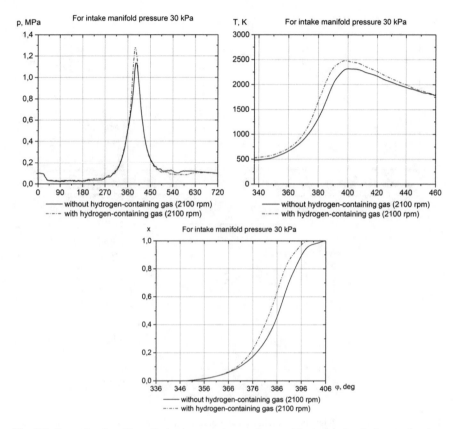

Fig. 9.8 Dependencies of in-cylinder pressure, temperature, and heat emission during combustion on the position of crankshaft with the addition of HCG

an increase in maximum in-cylinder pressure and temperature. In particular, at a speed of 1200 rpm when adding HCG, the maximum temperature increases in the cylinder by 100... 300 K, and the maximum pressure increases by 0.2... 0.5 MPa. The increase in parameters is 0.1... 0.3 MPa and 100... 200 K, respectively, at a speed of 2100 rpm.

To better understand the effect of adding HCG to the engine fresh charge on the process of combustion, a study of the dependences of pressure, temperature, and heat emission characteristics in the cylinder during combustion is carried out. An example of such dependences for a single engine mode (absolute inlet pressure of 30 kPa and speed of 2100 rpm) is shown in Fig. 9.8. In this mode, the HCG addition provides an increase in engine power from 1.32 to 1.98 kW (Fig. 9.7). As can be seen from the presented dependencies, the improvement of the engine fuel economy and energy efficiency performance when adding HCG is caused by a reduction in combustion time (7 degrees of crankshaft rotation for this mode). This reduction provides an increase in the maximum pressure in the cylinder from 1.14 MPa (which

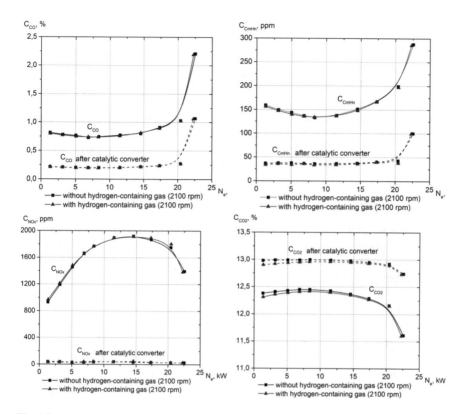

Fig. 9.9 Dependencies of concentrations of harmful substances in EG on the load of engine with the addition of HCG

without the addition of HCG is achieved at 392.5 deg. of position of the crankshaft) to 1.29 MPa (which when added HCG is achieved at 389.6 deg). The corresponding effect on the in-cylinder temperature has the parameters 2316 K at 400 deg. without the HCG addition and 2482 K at 397.9 deg. with the HCG addition.

The analysis of the influence of HCG addition on emissions of harmful substances was carried out according to their concentrations in EG of engine. An example of such analysis is presented in Fig. 9.9 for a speed of 2100 rpm. As can be seen from the presented dependences, the addition of HCG has almost no effect on the harmful substance concentration in EG. In this case, due to the HCG addition to the fresh charge and reduce the amount of fuel in the combustible mixture carbon content is decreased. This causes some decrease in CO and CO_2 concentrations. Reducing the duration of the combustion process, which occurs when adding HCG on the one hand, leads to a more complete fuel combustion and a reduction in the concentration of C_mH_n as a result and on the other to some increase in the concentration of NO_x. Since during the HCG addition to the fresh charge changes in the composition of the combustible mixture relative to the stoichiometric are absent, the effect on the catalytic converter efficiency is also absent. This ensures a minimum content of harmful substances in the EG when adding HCG.

The studies of generation of electrical power by TEG (Fig. 9.10) show that generated power increases with the engine speed and load increase. The obtained results are sufficiently correlated with theoretical research on a mathematical model. These results indicate that the additional battery efficient charging with electrical energy from the TEG is possible only in modes with high load. In these modes, the hydrogen generator energy consumption from the additional battery is absent that allows to maintain a sufficient level of charge of the additional battery at alternation of the engine operation modes in operating conditions.

The dependences of the engine parameters obtained in the experimental studies allowed to refine and verify the adequacy of the mathematical model used to further study of the HCG adding effect to the engine fresh charge on the vehicle emissions and fuel consumption while its motion is in a driving cycle. According to the results of comparison of experimental and calculated dependences of fuel consumption (Fig. 9.6), maximum in-cylinder pressure (Fig. 9.7), and TEG power (Fig. 9.10), it is established that developed model adequacy is at a sufficient level.

For further research, a mathematical description of the processes of the vehicle harmful substances emissions and fuel consumption during its conditional motion in the UNECE Regulation № 83–05 driving cycle is used. The simulation results show that when HCG is added to the fresh charge, the fuel consumption (Fig. 9.11) decreases from 708 to 694.5 g (by 1.9%), CO emissions (Fig. 9.12) decrease from 31.1 to 30 g (by 3.6%), C_mH_n emissions (Fig. 9.13) decrease from 1.286 to 1.248 g (by 3%), NO_x emissions (Fig. 9.14) decrease from 1.851 to 1.828 g (by 1.2%), and CO_2 emissions (Fig. 9.15) decrease from 1879 to 1837 g (by 2.2%). The decrease in

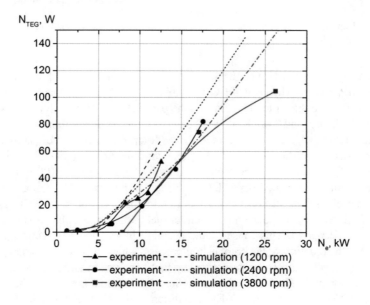

Fig. 9.10 Dependences of TEG power on speed and load of the engine

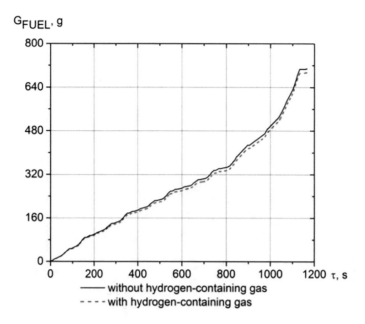

Fig. 9.11 Dependencies of fuel consumption on time of vehicle motion in the driving cycle with the addition of HCG

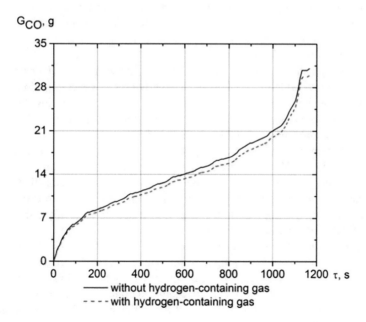

Fig. 9.12 Dependencies of CO emissions on time of vehicle motion in the driving cycle with the addition of HCG

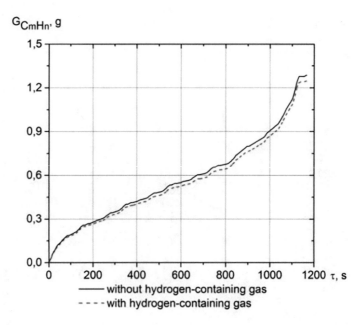

Fig. 9.13 Dependences of C_mH_n emissions on time of vehicle motion in the driving cycle with the addition of HCG

NO_x emissions despite the increase in their concentration with HCG addition to the fresh charge (Fig. 9.9) is associated with the effect on reducing air consumption (in accordance with the reduction of fuel consumption and constant fuel-air mixture composition, as shown in Fig. 9.6).

To determine the possibility of providing a sufficient level of charge of the additional battery, which is used to power the hydrogen generator, simulation of consumption and accumulation of electricity in this battery during the vehicle motion in the driving cycle. As the results show (Fig. 9.16), the hypothesis of the possibility of maintaining the degree of charge of the battery at a sufficient level due to the alternation of modes of consumption and accumulation of electricity in it is generally confirmed. However, there is a negative effect of not reaching the nominal level of battery charge due to insufficient power of the developed TEG. The investigated TEG power provides charging the additional battery to about 90% of nominal capacity. To improve this result, it is proposed to increase the nominal power of TEG to two and three times.

The power characteristic of the enlarged TEG is presented on Fig. 9.17. In this case, TEG power provides charging the additional battery appropriately to 94% and 96% of nominal capacity. This increases the efficiency of developed system under operating conditions. To achieve a more significant effect of the addition of HCG on improving environmental performance and fuel efficiency of the vehicular engine, further research is needed to define the optimal values of the HCG fraction in the fresh charge of the engine and the amount of thermoelectric modules in the TEG.

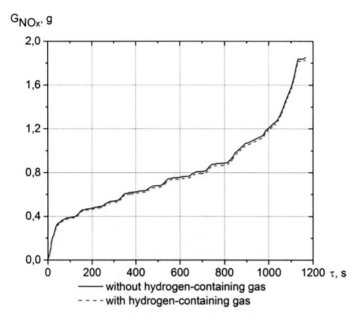

Fig. 9.14 Dependencies of NO$_x$ emissions on time of vehicle motion in the driving cycle with the addition of HCG

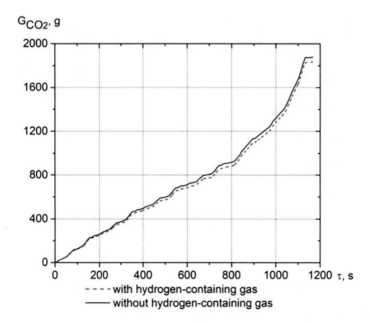

Fig. 9.15 Dependencies of CO$_2$ emissions on time of vehicle motion in the driving cycle with the addition of HCG

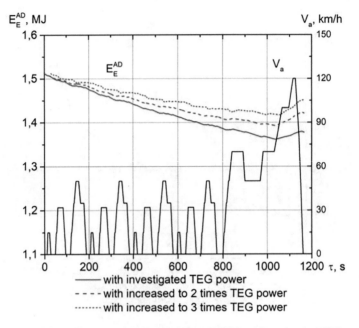

Fig. 9.16 Dependency of the accumulated energy in the additional battery using TEG on time of vehicle motion in the driving cycle with the addition of HCG

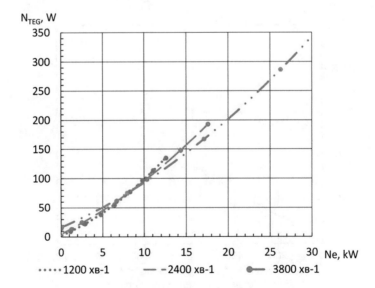

Fig. 9.17 Power characteristic of enlarged TEG

9.6 Conclusions

The addition of HCG is one of the most appropriate ways to reduce the vehicle emissions and fuel consumption as a result of reducing the consumption of carbon fuels of petroleum origin. To increase the efficiency of this method, it is important to provide a source of energy for the HCG production. One of the possible sources may be electrical energy obtained by recovering the unused engine EG heat energy provided by the thermoelectric converter.

The method of studying the influence of the HCG addition on the vehicle harmful substances emissions and fuel consumption provided a computational-experimental approach. The effect of adding HCG to the fresh charge of the engine on its energy efficiency and environmental performance, parameters of the operating process and combustion process, and the performance of TEG in certain modes of operation were determined experimentally. The impact of HCG addition to the engine fresh charge on fuel consumption and emissions, and efficiency of TEG to provide energy to the system of adding HCG during the vehicle motion in the driving cycle have been determined by calculation using the mathematical modeling.

The presented research results of the VW BBY engine show that the greatest effect of addition of HCG to a fresh charge on fuel consumption is achieved at the engine low load and speed. The decrease in fuel consumption is achieved in the range of 2.4–3.8% when adding HCG in the amount of 0.1–0.75% to the fresh charge due to the increase of the maximum in-cylinder pressure by 0.1–0.5 MPa and the maximum in-cylinder temperature by 100–300 K, which is caused by the reduction of combustion duration. Modeling of Skoda Fabia performance while its motion in the UNECE Regulation № 83–05 driving cycle shows that the studied fractions of adding HCG to the fresh charge of the engine improve fuel economy by 1.9% and environmental performance by 1.2–3.6% when the additional battery charge to power the hydrogen generator is provided with TEG at the level of 90%. However, the issues of further optimization of the parameters of the developed system require further research.

References

Anatychuk LI, Kuz RV (2012) Materials for vehicular thermoelectric generators. J Electron Mater 41(6):1778–1784

Artemenko RV, Tsiuman MP (2016) Evaluation the improvements of fuel economy and toxicity of ICE by utilization of heat energy in the thermoelectric generator. Herald of National Transport University. Ser Tech Sci Sci Tech Coll 1(34):21–28

Bari S, Mohammad-Esmaeil M (2010) Effect of H_2/O_2 addition in increasing the thermal efficiency of a diesel engine. Fuel 89(2):378–383

D'Andrea T, Henshaw P, Ting D, Sobiesiak A (2003) Investigating combustion enhancement and emissions reduction with the addition of $2H_2 + O_2$ to a SI engine. SAE technical papers. https://doi.org/10.4271/2003-32-0011

Dandrea T (2004) The addition of hydrogen to a gasoline-fuelled SI engine. Int J Hydrog Energy 29(14):1541–1552

Dmytrychenko MF, Gutarevych YF, Trifonov DM et al (2018) On the prospects of using thermo-electric generators with the cold start system of an internal combustion engine with a thermal battery. J Thermoelectr 4:49–54

Dyachenko V (2008) Internal combustion engines. Theory. NTU "KhPI", Kharkiv

Fairbanks J (2013) Automotive thermoelectric generators and HVAC. https://www.energy.gov/sites/prod/files/2014/03/f13/ace00e_fairbanks_2013_o.pdf. Accessed 6 Mar 2023

Gao M, Rowe DM (2002) Recent concepts in thermoelectric power generation. Paper presented at the twenty-first international conference on thermoelectrics, proceedings ICT'02, Long Beach, CA, USA, 29–29 August 2002

Gritsuk I, Volkov V, Mateichyk V et al (2017) The evaluation of vehicle fuel consumption and harmful emission using the heating system in a driving cycle. SAE Int J Fuels Lubr 10(1): 236–248

Gritsuk I, Mateichyk V, Tsiuman M et al (2018) Reducing harmful emissions of the vehicular engine by rapid after-start heating of the catalytic converter using thermal accumulator. SAE technical papers. https://doi.org/10.4271/2018-01-0784

Gutarevych Y, Shuba Y, Matijošius J et al (2018) Intensification of the combustion process in a gasoline engine by adding a hydrogen-containing gas. Int J Hydr Energy 43(33):16334–16343

Kuric I, Mateichyk V, Smieszek M et al (2018) The peculiarities of monitoring road vehicle performance and environmental impact. MATEC Web Conf ITEP'18 244:03003

Kuz R (2012) Moving vehicle parameters monitoring system. J Thermoelectr 14(4):86–90

Lakshmi DVN, Mishra TR, Das R, Mohapatra SS (2013) Effects of Brown gas performance and emission in a SI engine. Int J Sci Eng Res 4(12):170–173

Le AT, Nguyen DK, Tran TTH, Cao VT (2013) Improving performance and reducing pollution emissions of a carburetor gasoline engine by adding HHO gas into the intake manifold. SAE technical papers. https://doi.org/10.4271/2013-01-0104

Madyira DM, Harding WG (2014) Effect of HHO on four stroke petrol engine performance. SACCAM. http://hdl.handle.net/10210/13711

Mateichyk V, Saga M, Smieszek M et al (2020) Information and analytical system to monitor operating processes and environmental performance of vehicle propulsion systems. IOP Conf Ser Mater Sci Eng 776:012064

Tsiuman M (2019) Evaluation of fuel consumption and harmful substances emissions by vehicle with spark ignition engine under operation conditions with using of fuel containing ethanol. In: Boichenko S et al (eds) Selected aspects of providing the chemmotological reliability of the engineering. National Aviation University, Kyiv, pp 299–314

Tsiuman M, Artemenko R (2016) The peculiarities of heat losses simulation of automotive gasoline engine in operational conditions. Automob Trans 38:39–46

Tsiuman M, Gryshchuk O, Artemenko R et al (2019) Experimental study of thermoelectric generator for thermal energy utilization of internal combustion engine. Presented at SAKON'19, Transport (monograph) 17: 107–114, Poland, 18–21 September 2019

Tsiuman M, Mateichyk V, Smieszek M et al (2020) The system for adding hydrogen-containing gas to the air charge of the spark ignition engine using a thermoelectric generator. SAE technical papers. https://doi.org/10.4271/2020-01-2142

Vijayagopal R, Rousseau A (2015) Impact of TEGs on the fuel economy of conventional and hybrid vehicles. SAE technical papers. https://doi.org/10.4271/2015-01-1712

Wang S, Changwei J, Jian Z, Bo Z (2011) Improving the performance of a gasoline engine with the addition of hydrogen–oxygen mixtures. Int J Hydrog Energy 36(17):11164–11173

Wang S, Changwei J, Bo Z, Xiaolong L (2012) Performance of a hydroxygen-blended gasoline engine at different hydrogen volume fractions in the hydroxygen. Int J Hydrog Energy 37(17): 13209–13218

Zheng S, Fan W (2016) Simulations of TEG-based vehicle power system's impact on the fuel economy of hybrid and conventional vehicles. SAE technical papers. https://doi.org/10.4271/2016-01-0900

Index

A
Automobile road, vi, 123–142
Aviation infrastructure, 37

C
Computer modeling, 78, 92–96, 103
Computer systems, 75, 76, 88
Control, v, 13, 52, 148
Control system, 5, 14, 17, 23–25, 30, 33, 66, 71, 147

D
Driving cycle, 149, 156, 160–165
Drying, 29, 30, 72, 91–103, 108

E
Economical effect, 126
Electric vehicles, 59, 60
Energy efficiency, 109, 144, 145, 152, 153, 158, 165
Environmental effects, 125, 130, 142
Environmental impact assessment (EIA), v, 37–57
Environmental performance, 144, 145, 152, 153, 156, 162, 165
Environmental safety, v, vi, 48, 92, 93, 96–98, 107, 109–113, 144
Exploitation conditions, 5

F
Factor influences, 14–16, 25, 30, 32, 33
Flight critical and emergency situation, 5
Flight safety, 2, 4, 7, 9
Fluidized bed, 72, 91–103, 109, 114–117
Flying objects, 2, 4
Forecasting, 23, 30, 33

I
Information protection, vi, 76–79, 81, 88
Information technologies, 14

L
Lithium, 59–64, 66, 67, 69

M
Method, 2–5, 11, 14, 15, 20, 33, 42, 43, 45, 47, 56, 67, 73, 75–77, 79–83, 88, 92, 103, 109, 116, 119, 124, 130, 144, 147, 149, 150, 156, 165
Model, 3–5, 7, 11, 14–35, 76–84, 87, 88, 92–94, 100–102, 109, 110, 125, 126, 132, 133, 142, 146, 149, 151, 153, 154, 160
Monitoring, 2, 14, 21, 24, 33, 66, 75, 76, 78, 88, 107, 109
Multistage shelf dryers, 98, 101, 102

© The Editor(s) (if applicable) and The Author(s), under exclusive license to
Springer Nature Switzerland AG 2023
S. Boichenko et al. (eds.), *Sustainable Transport and Environmental Safety in Aviation*, Sustainable Aviation, https://doi.org/10.1007/978-3-031-34350-6

Index

N
Network, 28, 50, 51, 75–78, 88, 101

O
On-board hydrogen production system,
144–147, 149–151

P
Processes, 2, 15, 18, 20–25, 27, 29, 30, 38, 40,
42–47, 51, 52, 60–62, 66–70, 72, 75–85,
87, 88, 91, 92, 94, 96–98, 100, 103, 108,
109, 115, 116, 125–127, 149, 150, 152,
153, 155–160, 165
Projects of planned air transport activities, 38
Public private partnership, 124–126, 132–138,
141, 142
Public-private partnership, 123–142

R
Recycling, v, vi, 59–73, 108
Reliability, v, 13, 14, 17, 19–21, 24, 29, 30, 33,
76–78, 88

S
Safety information management, 1–11
Server, 76–88
Spent lithium-ion batteries, 59–73
Subway, 75–86, 88

T
Tariff for the use, 142
Technological trajectories, 109, 116, 118, 119
Technology map, 115–117
Technology readiness level, 103
Transport facilities (TFs), 38, 41–45, 50, 51, 56

U
Uncertainty, 13, 14, 16, 17, 22, 25, 27, 30, 33
Utilization, v, vi, 63, 67

V
Valuable metals, 61, 62, 66, 71–73
Vehicular engine, 144–165
Vortex granulators, vi, 109–117

Printed in the United States
by Baker & Taylor Publisher Services